Leaf protein and its by-products in human and animal nutrition

LEAF PROTEIN

and its by-products in human and animal nutrition

N. W. PIRIE

Second edition of
Leaf protein and other aspects of fodder fractionation

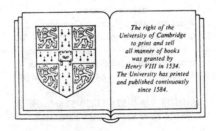

The right of the
University of Cambridge
to print and sell
all manner of books
was granted by
Henry VIII in 1534.
The University has printed
and published continuously
since 1584.

CAMBRIDGE UNIVERSITY PRESS

Cambridge

London   New York   New Rochelle

Melbourne   Sydney

CAMBRIDGE UNIVERSITY PRESS
Cambridge, New York, Melbourne, Madrid, Cape Town, Singapore, São Paulo

Cambridge University Press
The Edinburgh Building, Cambridge CB2 2RU, UK

Published in the United States of America by Cambridge University Press, New York

www.cambridge.org
Information on this title: www.cambridge.org/9780521330305

First edition (entitled *Leaf protein and other aspects of fodder fractionation*)
published 1978
Second edition published 1987

A catalogue record for this publication is available from the British Library

Library of Congress Cataloguing in Publication data
Pirie, N. W. (Norman Wingate), 1907–
Leaf protein and its by-products in human and
animal nutrition.

Rev. ed. of: Leaf protein and other aspects of
fodder fractionation. 1st ed. 1978.
Bibliography
Includes indexes.
1. Plant proteins. 2. Leaves–Composition.
3. Forage plants–Composition. I. Pirie, N. W.
(Norman Wingate), 1907–   . Leaf protein and other
aspects of fodder fractionation. II. Title.
TP453.P7P53   1987   641.3'3   86-24416

ISBN-13 978-0-521-33030-5 hardback
ISBN-10 0-521-33030-0 hardback

Transferred to digital printing 2005

Contents

Introduction

Protein sources dominated discussion on world food problems 30 years ago. That exclusive emphasis caused a natural reaction, and there was a period of preoccupation with energy needs. Those who, early in this century, had been taught of the 'protein-sparing action of the carbohydrates' were puzzled by such exclusive, or either/or, opinions. Unfortunately, the sensible middle-of-the-road outlook was epitomised in the foolish phrase 'protein–energy malnutrition' or PEM. When both protein and energy are scarce, people are simply hungry or half-starved though they may, if the concept PEM is treated logically, be supplied with such dietary components as minerals and vitamins. The simple words *hungry* and *starved* seem preferable to any type of scientific euphemism. Sometimes, all that is needed to remedy matters is an increased supply of the conventional foods of a region, and its equitable distribution both between and within families (Pirie, 1982). Equality of distribution raises political and social issues. Because of the complexity of these issues, it is fortunate that they are not the business of those whose primary concern is the food supply. But it is our business to try to ensure that agriculture produces nutrients in the amounts and ratios needed in each region (Pirie, 1981*a*; Swaminathan, 1981; Pinstrup-Andersen, 1982).

This book is concerned with the merits of one potential source of edible protein. It is therefore not a suitable place to discuss the varied opinions which have been expressed by individuals and international committees about the correct ratio between protein and energy in the diets of apparently well-nourished people. I have discussed that issue in several other publications (e.g. Pirie, 1984*b*, *c*). For a less biassed account, three papers in a recent conference organised by the Rank Prize Funds (Blaxter & Waterlow, 1985)

vii

should be consulted. Some confusion is introduced into the whole subject by misconceptions about the significance of nitrogen (N) balance measurements. Clearly, anyone not in balance, i.e. who is excreting more N than is being absorbed from food, is being depleted and must in time show signs of malnutrition. There is, however, no logical or experimental basis for the assumption that all is well with N metabolism as soon as intake equals output. That assumption confuses a necessary with a sufficient condition. Furthermore, it is not as easy as is often assumed to measure the point of balance. A committee of the United Nations University (1980) concluded that our apparent protein requirement increases as fewer assumptions are made about diet and experimental technique. Suggested values for the amount of protein needed per day have consequently been increasing (e.g. Young & Bier, 1981; Gersovitz *et al.*, 1982). We seem to be slowly returning to the old, and easily remembered, value of 1 g of protein kg^{-1} of body weight. Without a protein concentrate in the diet it would be impossible to approach that value if banana, cassava or sago were the main energy sources, and difficult if the energy sources were potato or rice. Although it has recently been fashionable to dispute statements such as that, disagreement is clearly not taken seriously by those responsible for national policies. Otherwise, less effort would be expended on cultivating seed legumes, fishing, milk production, and poultry husbandry. Work on protein sources other than these is therefore justified.

Leaves are potentially the most abundant source of edible protein (Pirie, 1975*b*, 1981*b*). The best and simplest way to exploit this potentiality is to eat more leafy vegetables in the normal manner (Pirie, 1985*a*). However, the human gut has a limited capacity (which few people approach) to cope with whole leaf. In many parts of the world it would therefore be advantageous to make extracted leaf protein (LP) because:

1. Leaves are the site of protein synthesis and there are losses when protein is translocated to seeds or tubers.
2. Suitable leaf crops maintain a photosynthetically active green cover on the ground throughout the period during which growth is possible. Yields are therefore greater than those from crops which occupy ground while merely ripening. Perennial leaf crops protect ground from erosion.
3. Ruminants make admirable use of nonarable land, but they

convert only 10 to 25% of the protein in their fodder into human food. From cultivated plants, 50 to 65% of the protein can be extracted, and the unextracted protein is still available for ruminants.

4. The processes of extracting and separating LP disintegrate the leaf and remove toxic or ill-flavoured components. Species normally rejected as human or animal food can therefore be used.

5. Forage from which LP will be made is harvested when young. The fibre is therefore less lignified than when a crop is taken for hay. Furthermore, the crop is not at risk from pests and diseases for so long.

6. The process of extraction removes much of the water from the fibrous residue. When ensiled, there is therefore less drip; when conserved by drying, less fuel is needed. Field wilting is a common technique for achieving these results. But 30% of a wilted crop may not be gathered up. It is easy to collect 95% of a crop which is cut and harvested in one operation.

7. If made with reasonable care, LP has better nutritive value than the usual seed proteins; it is as good as fish or meat, but not egg or milk. Like other foods containing unsaturated fats, it is damaged by inept handling.

8. Although people with European or North American prejudices find the appearance of LP unusual on first contact, it is readily accepted by adults and children when intelligently presented.

9. The technique of extraction is simple; equipment has been designed, and is being constantly improved, for production on the domestic, village and commercial scale.

10. Several countries which are so wealthy as to have little need for LP as a human food, depend on imported groundnuts and soya beans for pigs and poultry. LP could largely replace these imports.

The validity of these points is now gaining acceptance, especially among those who are concerned with feeding animals. This is partly because increases in the cost of oil necessitate less waste of fuel, and partly because people who heard of fodder fractionation when young are now reaching influential positions. The present degree of interest is shown by the need for a second edition of this book and by the appearance of two other books (Costes, 1981; Telek & Graham, 1983). Increasing international interest is shown by the conferences on LP, in India in 1982 (N. Singh, 1984) and Japan in 1985 (Tasaki, 1986), which have, after a long interval, followed the conference in India in 1970 (Pirie, 1971*a*). A conference in Italy is planned for 1989.

Abstracts of about 500 papers dealing with LP and its use as human food, as feed for nonruminant animals such as pigs and poultry, and with the use, as ruminant fodder, of the residue from which LP has been extracted, which have appeared since the first edition of this book, have been collected and circulated in nine numbers of a *Leaf Protein Newsletter* (Matai, 1984). We hope to continue this service.

National and international organisations concerned with research on food have shown little interest in the use of LP as a human food. Charitable organisations, notably 'Find Your Feet', have been mainly responsible for the progress which has been made. Elsewhere (e.g. Pirie, 1976*a*), I have discussed some reasons for official reluctance to admit that, with the present rate of population increase, reliance on a steady increase in agricultural productivity may be mistaken, and new methods of food production may be needed. I have also censured the tendency, in some quarters, to try to solve the food problem by changing the assumptions made about our food requirements. Both points have recently been forcefully emphasised by Miller (1983).

Though welcome, this widespread acceptance of the idea of fodder fractionation encourages the assumption that we now have adequate knowledge about what should be grown for fractionation, how it should be fractionated, and how the products should be used. This is far from the case. The amount of research that has been done is trivial compared to the amount that has been, and is being, done on projects of comparable complexity and smaller potential yield, such as the cultivation of microorganisms or the processing of fish. Some of the more obvious topics on which much more research is needed are outlined in this book. It is well to remember that haymaking and ploughing are ancient arts on which useful research is still being done.

An anomalous feature of agricultural research is that, while magnificent work is done on increasing the efficiency of primary photosynthetic production, on ensuring that a plant's roots are adequately supplied with mineral nutrients, and on selecting varieties with increased Harvest Index (the ratio of total dry matter to conventionally useful DM), little attention is paid to making optimal use of crops which already have Harvest Indexes approaching 1, i.e. leafy crops.

Terminology is important in every subject. Agreement about it is convenient, but it is more important to ensure that it is neither ambiguous nor misleading. There is no ambiguity about the equivalent words 'juice' and 'extract'. The coagulum separated from leaf juice, but not further fractionated, is called leaf protein (LP) in this book. It is often called leaf protein concentrate (LPC) by others, and some recent publications have called it leaf nutrient. The last name is suggested because of the valuable presence of β carotene (pro-vitamin A) and traces of other vitamins. But it will cause confusion in indexes because leaf nutrient already means something which is sprayed on to the leaves of a growing plant. As normally made, LP is a mixture of many proteins; the intrusion of the word *concentrate* suggests that a product is being referred to which has a greater content of true protein than some parent substance that would be called leaf protein. It can be argued that it should be called leaf lipoprotein because it contains 30 to 40% of non-protein material – most of it lipid. That would imply that most of the lipid was attached to, rather than merely mixed with, most of the protein. This is probably not the case. When LP is fractionated, the usual products are a green mixture of chloroplasts, their fragments, and fragments of fibre, etc., not removed by straining the juice. Here this is loosely called 'chloroplast' protein. The other, paler, product is equally loosely called 'cytoplasm' protein although much of it was originally in the chloroplasts. It is important not to give the residue from which protein has been extracted a misleading name. Here it is called fibre – which is brief. It could be called extracted residue. The most misleading name for it is pressed crop: that perpetuates the widespread illusion that pressing is the important feature of fractionation. Here, juice from which protein has been removed is called 'whey'; that is brief and the metaphor is obvious. It is sometimes longwindedly, but not misleadingly, called brown juice or deproteinised juice.

In some publications it is not clear whether the yields given are for moist press-cake of LP, dry LP, or the true protein component of the LP. Here, wherever the contrary is not stated, the yields given are for 100% protein, and they are usually calculated by multiplying the N content by six.

Abbreviations

ARC   Agricultural Research Council; now Agricultural and
      Food Research Council, AFRC
cm   centimetre
DGLV dark green leafy vegetable
DM   dry matter
FAO   Food and Agriculture Organisation of the United Nations
g   gram
g   normal gravitational acceleration, 9.8 m s^{-2}
ha   hectare; ha^{-1} = per hectare (the same convention is used
      for other quantities)
HP   horse-power = 746 watts (W)
IBP   International Biological Program; this has now ended
J   joule = 1 watt second = 0.239 small calories
kg   kilogram (kgf = kilograms force)
km   kilometre
kPa   kilo Pascal = 1/98 kgf cm^{-2}
kWh   kilowatt hour, often loosely called a unit of electricity
l   litre
LP   leaf protein (unfractionated)
m   metre
mg   milligram
ml   millilitre
mm   millimetre
MJ   megajoule = 1 million joules = 239 kilocalories
M£   1 million pounds sterling
µg   millionth of a gram
µm   millionth of a metre

N	nitrogen (except when gaseous N_2 is referred to)
RNA	ribonucleic acid
RuBP	ribulose-1,5-bisphosphate carboxylase-oxidase = rubisco = fraction 1 protein
s	second
t	ton or tonne (tf = tons force)
TCA	trichloroacetic acid
UNICEF	United Nations Children's Emergency Fund
UNU	United Nations University
USDA	United States Department of Agriculture
WHO	World Health Organisation

1

Historical and anatomical background

It is seldom possible to date precisely the beginning of a new line of research. Preliminary vague hints, coming from folk medicine, primitive technology or some such sources, are as a rule slowly integrated into a new fabric. Leaf protein is different. The date of the first publication on it is 1773. The chemical category now known as 'protein' was not recognised at that time: the word itself was not coined until 1838 when Berzelius used it in a letter to Mulder. Presumably because of some similarity in texture and sheen, wheat gluten was called *mien chin*, literally muscle of wheat, centuries ago in China. When Beccari studied gluten in 1728 (published in 1745) he emphasised a further relationship between gluten and animal products – they stank similarly when putrid or heated. The same criterion led Rouelle (1773) to call the material he isolated from leaves *matière glutineuse ou végéto-animale*. The 'smell of burnt feathers' retained a place in textbooks, as a means of protein recognition, for many years.

About 1785, Berthollet found N in this group of substances, the colour test known as the xanthoproteic reaction was described by Fourcroy and Vauquelin in 1800, and Fourcroy stressed the importance of N in the nutrition of plants and animals in about 1806. These observations were reasonably factual. Then, as was his way, Liebig confused the issue by using his imagination rather than his undoubted experimental skill. He decided that there were only four proteins and asserted that the curds separating from boiled vegetable juices were indistinguishable from those separating from such animal fluids as blood and egg white. As Berzelius remarked in 1842: 'This easy kind of physiological chemistry is created at the writing desk, and is the more dangerous, the more genius goes into

1

its execution.' The individuality of proteins was not established till many years later.

The brothers Rouelle, though very unlike in temperament and appearance, are often confused. Their portraits have even been misassigned. Guillaume François (1703–70) was an enthusiastic and excitable teacher. His manner of teaching was described as '...*souvent incorrecte et familière, mais toujours animée et pittoresque,...*' and he enlivened proceedings on some occasions by partly undressing, and by explosions. Even when young he was megalomaniac, accusing many contemporaries of plagiarism, using such phrases as '*Ecoutez-moi! car je suis le seul qui puisse vous démontrer ses verités*'; in 1756 he claimed to have a secret weapon with which he could destroy London and the British navy. Patriotic feeling was so strong that he refused a tempting offer from a London publisher; his lectures remained unpublished. Lavoisier and Rousseau were among his distinguished pupils, and it is after him that the rue Rouelle in Paris is named. Because of increasing eccentricity he resigned his professorship at the Jardin du Roi (now Jardin des Plantes) in 1768. At his request, his brother, Hilaire Marin (1718–79), succeeded him – but only as demonstrator.

By contrast, H. M. Rouelle was neat and tactful; an experimenter rather than a forceful teacher. He did not share his brother's chauvinism, but became a corresponding member of the Royal Society of Arts (London) on 24 October 1770. There is no record of him playing an active role in the Society. J. d'Arcet wrote an *éloge* in *Observations sur la Physique* (**16**, 165, 1780). He is less often mentioned than his brother, and gets less space when mentioned in biographical dictionaries in spite of a distinguished research record. He found potassium in 'cream of tartar' and went on to isolate tartaric acid from unfermented grape juice and to make tartarates of several metals. He found formic acid in ants, and made urea, or possibly the hydrate of urea and sodium chloride, from urine. This was not a novelty: Boerhaave probably made it in 1729. He confirmed the presence of iron in blood. This hardly needed confirmation, for the removal of rust stains from blood-soiled clothing must have been a familiar problem. He extended his brother's studies on various oils and resins. In this work he pioneered the use of solvents in sequence. At that time, starch and fat were considered the nutritionally important components of foodstuffs. Rouelle, having begun to recognise proteins as a chemical category, stressed their nutritional importance. Clearly, he helped to establish the science of biochemistry: he has been called its father. However, he had not quite shaken off the older outlook. For example, he did not completely reject the possibility of alchemical transformations, and argued that, although we could not make an animal or a plant, it did not follow that a metal, without structure or parts and therefore presumably lifeless, could not be made. The argument that the characteristic feature of an organism was its heterogeneity had already been used by Jean Rey in 1630 (cf. Pirie, 1964a).

Rouelle (1773) published two papers on leaf protein. The second contains so much information that it can be taken as the origin of our subject. Its title, '*Sur les Fécules ou parties vertes des Plantes, & sur la matière glutineuse ou végéto-animale*', shows his recognition that plant and animal products are similar. He pounded leaves of several species in a marble mortar, pressed out the juice and heated it. A green coagulum, which he could decolourise by washing with alcohol, separated at about the temperature at which he could no longer keep his finger in the hot juice.* He filtered off that coagulum and got a pale coagulum on further heating. Although this paper is now often referred to, and since about 1952 the correct initials are usually given to its author, it seems to have been seldom read. Patent Office inspectors are particularly remiss and allow patents that cover points clearly established in 1773! I therefore give a free translation of the paper in an appendix to this book. Rouelle's work was extended by Vauquelin and Fourcroy (1789) a few years later.

In spite of the defects and tediousness of contemporary methods for measuring N, Boussingault concluded in 1839 that atmospheric N_2 (dinitrogen) and food N were not interchanged during animal metabolism. The generalisation that N_2 is little, if at all, used by animals, or made by them from their food, is still accepted by most scientists though there are a few puzzling apparent exceptions. Mulder attached particular importance to plant proteins as the source from which animals derive their protein and other nitro-

---

* Farenheit made thermometers and developed the idea of a temperature scale in 1742: Martel described the Centigrade scale in the same year. This scale is sometimes, absurdly, called the Celsius scale although Celsius had it upside down with water boiling at 0°. Meteorologists adopted thermometers quickly – chemists did not, for example: thermometers are not mentioned in Macquer's (1766) *Dictionnaire de chymie*.

genous substances. An animal, as Johnson phrased it in 1867, 'moulds over these vegetable principles into the fibrine, albumin and casein of its muscle and other tissues, of its blood, milk and other secretions'. As a step towards assessing their value as fodder, forages and other animal feeding stuffs were analysed in many laboratories – notably at Rothamsted by Lawes and Gilbert.

Beddoes suggested in 1792 that leafy material should be made into human food (Levere, 1984). Like most of Beddoes' suggestions, this was ridiculed : he figures in scientific literature mainly because he started Davy on his career. The suggestion was probably made again during the 19th century because Lawes (1885) remarked 'It might be possible by some chemical process to produce from grass a nutritious substance which a man could use as food, but food so extracted would be far more costly than as it existed in the grass, and no one would think of preparing such a food for oxen or sheep, as their machinery is quite competent to separate the nutritious from the indigestible portion of the food.' Lawes' comment on ruminants is justified: it is fortunate that his comment on the extraction of human food was not widely known when work on LP extraction started at Rothamsted!

Winterstein (1901) extracted protein from dried, ground leaves with dilute alkali. Sustained work started 20 years later when Osborne temporarily forsook his studies on the seed proteins. Osborne & Wakeman (1920) and Osborne *et al.* (1921) made preparations from spinach (*Spinacia oleracea*) and lucerne (*Medicago sativa*) by pulping the fresh leaf, removing the coarser particles by centrifuging or filtering, and coagulating with alcohol.

Osborne (1924) was well aware of the importance of this work, but he was distressed by the properties of the material. As he put it: 'Our present meagre knowledge of the protein constituents of living plants is chiefly due to the difficulties encountered in separating the contents of the cells from the enveloping walls. Attempts to grind the fresh leaf and extract the contents of the cells with water result in mixtures that cannot be filtered clear, and consequently appear to present no opportunity to obtain the protein in a state fit for chemical examination.' As Vickery (1956) put it: '...when Chibnall came to the laboratory in 1922 for a two-year period, Osborne was happy to turn this difficult problem over to him...' Chibnall had already worked with extracts from cabbage (*Brassica oleracea*) and runner bean (*Phaseolus vulgaris*). One chapter in the book he wrote in 1939, *Protein metabolism in the plant*, describes his work.

The work of Chibnall and Osborne established LP as a reasonable material for biochemical investigation. A steady flow of papers on techniques of extraction in the laboratory soon started e.g. Davies (1926); Lugg (1932, 1939); Kiesel *et al.* (1934); Yemm (1937); Foreman (1938); Crook (1946); Crook & Holden (1948); Holden & Tracey (1948); Bryant & Fowden (1959); and Festenstein (1961). Because leaves from different species, and of differing age and nutritional status, were studied the conclusions reached in these papers are not identical. However, the general conclusion was that the younger the leaf and the greater its protein and water content, the greater the percentage extraction of protein. Also, thorough subdivision or rubbing, and the maintenance of alkaline conditions during the extraction, are advantageous.

During the past 30 to 35 years the total amount of biochemical research has increased enormously. The amount of research on plants has not increased correspondingly, and that research tends to be concerned with alkaloids, pigments and the components of seeds and tubers. Nevertheless, much more work has been done on proteins in leaves than it would be useful to describe in detail in a book primarily concerned with practical problems arising when an edible protein is being extracted in bulk. Early academic work on leaf proteins is described in several reviews, e.g. Vickery (1945), Pirie (1955, 1959a), and various specialised aspects of the subject are covered regularly in review journals. It may, nevertheless, be useful to give a brief survey of the subject so as to explain the reasons for some methods and precautions adopted during the production of LP.

Even casual study of pieces of leaf under the microscope shows that the green colour is concentrated in chloroplasts: the colour of leaf extracts shows that chloroplasts, or their fragments, are being extracted. Methods for separating them from the other types of protein will be discussed later (p. 66). Here, all that need be said is that there is much research on their isolation in physiologically active forms. Those from some species are extremely fragile, while from others they are robust enough to withstand the disintegration of the leaf during digestion by *Clostridium roseum* (White *et al.*, (1948). Morris & Hall (1982) comment on the unusual stability of chloroplasts from quinoa (*Chenopodium quinoa*). The durability of chloroplasts depends on the environment into which they are put. Extreme examples are the prolonged photosynthetic activity of chloroplasts introduced artificially into mammalian cells (Nass, 1969), and the frequent survival of algal chloroplasts in molluscs (Taylor, 1968; Trench *et al.*, 1973; Hinde & Smith, 1975) and other invertebrates (McNeil & Smith, 1982). It would be worthwhile studying methods for increasing fragility: for example, by electric shock in the manner which is effective with erythrocytes (Zimmermann *et al.*, 1975). If chloroplasts were more completely broken, less LP would probably be retained by the fibre. Increased extraction of LP when leaves are pulped with added water may be partly caused by chloroplast damage from osmotic shock.

When juice is pressed out from leaf pulp, the extent to which chloroplasts are extracted depends on their integrity and on the compaction of the leaf fibre. Typical chloroplasts are 5 μm wide and 2 μm thick. It is therefore advantageous to choose species with fragile chloroplasts, to keep the layer of pulp thin, and to express gently as much juice as possible so as to avoid pressing the fibre into a relatively impermeable mat. These points are important because much, probably most, of the protein is in the chloroplasts initially.

The smaller cellular components, such as mitochondria and ribosomes, also contain protein and they are less likely to be held back by compacted fibre. Furthermore, they disintegrate readily. This is doubly advantageous; it prevents loss and it diminishes the amount of nucleic acid that separates along with the LP. Extracts from young leaves may contain 2 or 3 g of ribosomes l^{-1}, and these contain 20 to 30% of RNA (Pirie, 1950). Provided the diet does not contain much RNA from other sources, the quantity that would be eaten in LP would be harmless, but, especially for people with a tendency to gout, all sources of nucleic acid must be watched. Leaf extracts contain ribonuclease so, as will be explained later (pp. 17, 51), nucleic acid is easily degraded in them.

Many enzymes are present in leaf extracts; their molecules are mostly too small to be retained by a packed mat of fibre to a serious extent. The dominant component is ribulose-1, 5-bisphosphate carboxylase-oxidase, hereafter called RuBP (also often called rubisco or fraction 1 protein) which is an important component of the photosynthetic mechanism. Preparations of this enzyme from different species have similar amino acid compositions; that similarity, together with the large number of different enzymes making up LP, probably explains the constancy of its composition (p. 62). Problems arise in handling leaf extracts because of the presence of enzymes that catalyse the oxidation of phenolic substances and produce tanning agents that coagulate protein and diminish its digestibility.

However thoroughly leaves are ground, some N remains attached to the fibre. Part of this is probably protein that was initially soluble but gets attached to the fibre as a result of the stresses (perhaps momentary local heating) during grinding (Bawden & Pirie, 1944; Crook, 1946). The fibre residue is always bright green, and therefore has some chloroplasts or their fragments entangled in it. Holden & Tracey (1950) measured the ratio of chlorophyll to N in preparations of chloroplasts and extracted fibre, and concluded that 80% of the N could be accounted for thus. They suggested that the change in texture, which takes place when fibre is incubated with proteolytic enzymes, could be the result of removal of non-chloroplast protein from cell walls. This suggestion gains strong support from the recognition of hydroxy proline (a characteristic component of cell walls elsewhere) in the fibrous residue (Clarke & Ellinger, 1967; Jennings & Watt, 1967). Laird *et al.* (1976), working with young lucerne leaves, found that the amount of N which remained insoluble even after extraction with a mixture of phenol, acetic acid and water, was more than one-third of the amount present in readily extractable protein.

Those who are interested in extracting protein in bulk are usually satisfied if 50 to 70% of the protein N is being extracted, and choose species from the young leaves of which that degree of extraction is possible. The precise nature of the unextracted N, or N extracted with difficulty, is, however, of great interest to those wishing to use the residue as cattle fodder.

2

Prelude to production

Beddoes in 1792 and Lawes (1885) suggested the possibility of making human food from inedible leaves: Ereky (1927) took out the first patent on the idea. This was the forerunner of patents by many others for methods for making food for nonruminants from material that could otherwise be used as ruminant food only. Ereky, who was Minister of Development during the Horthy dictatorship in Hungary, tried to get an article on feeding ducks and pigs on his product published in a British journal in 1924 and 1925. He was secretive about his process, because of the impending patent, and the article seems not to have been published. Three-quarters of his patent specification is devoted to confused and largely erroneous statements about changes undergone by leaves during autolysis and drying. The machine was designed to cut, rather than crush, forage between knives set on opposed faces of a pair of truncated cones. The forage was carried round (as in the Hollanders used in paper making) in a stream of water which was recycled after separating the LP by an unspecified method. In spite of his official position, but not unexpectedly considering the inadequacy of his machine, he was unable to make enough LP for the prolonged feeding trials that British editors asked for. Ereky seems to have been unaware that what he claimed had already been done by Rouelle, Osborne and Chibnall, but he deserves attention because he had a clear idea of the advantages of fractionating forage.

The defects in Ereky's proposed machine were recognised by Goodall whose patent (1936) covered the use of three-roll mills, expellers and belt presses. He stressed that rubbing rather than cutting was essential as a first step in getting juice out of leaves, and that total disintegration into a purée was unnecessary and complicated subsequent handling. He was primarily interested in

8

the medicinal merits of his product because it contained 'chlorophyl, vitamins, and ferments', but he also attached importance to the partly dewatered fibre because it could be economically dried in a current of unheated air to produce a winter feed for cattle. The use of juice from grass as a medicinal product was also patented by Schnabel (1938) who had for many years a somewhat mystical enthusiasm for pills made from protein-rich grass harvested before the formation of the first joint.

After this digression into pharmacology, interest reverted to the extraction of food from leaves. Slade (1937) stressed the productivity of grassland and argued that, if it should be necessary to produce more of our own food in Britain, it would be better to fertilise grassland heavily, extract LP as a human food, and feed ruminants on the residue, rather than to plough up grassland. The proposition was included in a patent (Slade *et al.*, 1939) which also covered many of the processes that had been used by Osborne and Chibnall, and that were regularly used in the separation of plant viruses from leaf extracts. There was more novelty in the suggestions that the protein curd should be compressed anaerobically and allowed to mature into a 'cheese', and that yeasts should be cultivated on the filtrate from the curd. Many patents have been allowed since 1939 although it is by no means clear, in the light of all the early academic work, and of the patents that have now lapsed, what points of any significance are being covered.

By 1939 the position had been fully defined: the quantity and extractability of protein in a few leaves was known, the advantages of extracting it were recognised, some methods for doing this were known, and so were the bulk properties of the protein. All that remained was to find out whether the production of leaf protein in bulk was practicable and whether the protein would prove to be as useful as theory suggested. I had started making plant virus preparations in 1934 and so gained experience in handling the normal proteins of the leaf. On the outbreak of war in 1939, many meetings were held in laboratories in Cambridge and elsewhere to discuss the ways in which scientific skill could be best used. Those of us who were interested in nutrition, remembering the effects of blockade in the 1914–18 war, suggested various projects for improving food production, with particular emphasis on protein. I suggested work designed to see whether new academic knowledge made bulk production of leaf protein feasible. Nothing happened immediately, but in the general ferment that followed the 'fall of France' in June 1940, I was asked to help the Food Investigation Board and Imperial Chemical Industries in a joint study of the problem. It seemed clear that there was little need for more work in the laboratory; instead, large-scale machinery had to be found or designed.

During 1940 and 1941, through the friendly cooperation of manufacturers and industrial users, many different hammer mills, screw expellers, sugar cane rolls, ball mills, rod mills, edge- and end-runner mills and dough-breakers (incorporators or Pfleiderers) were tried; all proved usable for making a satisfactory pulp, but they were all, for various reasons, unsatisfactory. Stamping mills of the type used to disintegrate ores were not then tried. A few later experiments (Pirie, 1953) suggested that they had potentialities, but more recent trials have convinced me that pounding without some form of rubbing or shear is not likely to be effective.

It was often difficult to persuade manufacturers of various pulping machines, tested for such a short time that an equilibrium temperature was not reached, that frictional heating was excessive. But it is obvious from the value of the mechanical equivalent of heat that, if 30 HP is consumed in making 1 t hour^{-1} of pulp containing 85% of water, the temperature of the issuing pulp will be raised by nearly 30 °C. With such power consumption, protein would therefore be coagulated *in situ* on a hot day unless one resorted to the further complication of cooling the pulper. There are therefore reasons, other than saving energy, for avoiding useless friction. With proper disposition of beater arms in units with the general character of a hammer mill, it is possible to make 1 t of pulp hour^{-1} using as little as 10 HP. Addy *et al.* (1983) describe experiments on thin layers of leaf which suggest that it should be possible to use only a sixth as much power as that.

These tests on large-scale equipment, and some pilot-plant production of LP for palatability trials, started with ample encouragement from the relevant official bodies. There was a formal lunch in Cambridge in October 1940, attended by the Minister of Food, the Regional Commissioner, the Divisional Food Officer and others, at which LP soup, whale casserole, and other novelties were served. Questions were asked about LP in the House of Commons (e.g. Hansard 15/10/41, column 1370). By the end of 1941, euphoria had waned and the familiar official distrust of every unconventional proposal reasserted itself. There was a hilarious period in which I was told, at the same time and sometimes by the same people, that the idea of making LP as a human food could not possibly be practical, and that it was so important that I must be more secretive and not explain, to those whose pulpers I was testing, what was intended, lest the Germans should get hold of an important idea. The generous supply of food from the USA under 'Lend-Lease' finally made it obvious that there would be no need for British self-sufficiency in wartime.

Although government support for work on LP ended, considerable interest was still taken in the idea. In Britain, LP was discussed at several scientific meetings (e.g. Pirie, 1942a, b) and it was often mentioned in the popular press. Guha (1960) used LP from water hyacinth (*Eichhornia crassipes*) and other species during the Bengal famine in 1943, and Australians in a Japanese prisoner of war camp crushed 300 t of grass, rhododendron and passion fruit leaf between rollers so as to make an extract containing vitamins and protein (Smith & Woodruff, 1951). LP was used at the same time in the USSR as a source of vitamin A in human food, and as a protein supplement for calves, hens and pigs (Zubrilin, 1963, in Pirie, 1963). A return to the Winterstein (1901) method of drying and extracting with alkali was advocated in the USA (Sullivan, 1943, 1944). This process, as would be expected, has never been used on a large scale anywhere. In both Nebraska and California (Bickoff *et al.*, 1947), LP was made by pulping fresh forage.

By 1948, interest in LP seemed to have disappeared in India, the USA and USSR. In Britain, the end of 'Lend-Lease' food supplies directed attention to the need for increased home production. There was, at the same time, more general awareness of the poor nutritional state of most of the world's inhabitants – especially those living in the wet tropics. Consequently, work on LP started again at Rothamsted and the Grassland Research Station (then near Stratford-upon-Avon), supported by a grant from the ARC. Work at the Grassland Research Station was discontinued in 1953; work on LP at Rothamsted never formed part of the general program of the station but depended on special grants. Later improvements in technique and equipment depended mainly on a five-year grant from the Rockefeller Foundation in 1958, and a three-year grant from the Wolfson Foundation in 1965.

The early trials with large-scale equipment showed that hammer mills would pulp leaves satisfactorily if clogging could be prevented. In a conventional hammer mill, the charge remains inside until comminuted sufficiently to pass through holes in part of the casing. The dough that is made from pulped leaves clogs holes small enough to ensure adequate subdivision. This defect can be overcome by flushing the pulp through with water. That method has been advocated (Chayen, 1959; Chayen *et al.*, 1961) with the misleading designation 'impulse rendering', although it produces an inconveniently dilute extract. During discussion with Mr William Christy (of Messrs Christy & Norris, Chelmsford) in 1940, the idea evolved of making a hammer mill with a cylindrical pulping chamber two or three times as long as it was wide. This would have the intake at one end of the cylinder and discharge at the other, without any grid or screen. It would therefore be uncloggable: whatever went in would come out, whether properly pulped or not. To get the correct degree of pulping, an arrangement of beaters would be chosen that moved the charge through the pulper at a suitable rate. When work restarted in 1948, a 'coir sifter', normally used to separate coconut husk from fibre, was modified and used to pulp 1 to 2 t (fresh weight) of crop hour^{-1}. During the next 20 years many more modifications were made (e.g. Davys & Pirie, 1960) so that latterly the ancestral role of the coir sifter was no longer obvious. There is no need to describe them because pulpers such as these are not likely to be used to make LP in the future. However, the same principle underlies the design of the 100 kg hour^{-1} pulper, designed for the International Biological Program (IBP) (Davys & Pirie, 1969), which is still used in several institutes.

Trials on a laboratory scale showed that 80 to 90% of the amount of juice which can be expressed at any pressure, is expressible in 5 to 10 s at 1.5 to 3 kgf cm^{-2} (150 to 300 kPa) if the layer of pulp is less than 2 cm thick. Obviously, power is wasted if there is relative motion between the charge and filtering surface while under pressure. Several presses meeting these specifications were made. In the first a ram pressed down on pulp carried on a hinged loop of a perforated conveyor; this moved in steps between strokes of the ram. At the instigation of the National Research Development Corporation this was patented (Pirie *et al.*, 1958) although it had already been superseded by a press in which the pulp was carried on a rotating perforated table (Davys & Pirie, 1960). After considering more fully the principles on which successful expression of protein-rich juice depends (Pirie, 1959*b*), we (Davys & Pirie, 1965) made a reasonably satisfactory press in which pulp is pressed against a perforated pulley by a tensioned flexible belt. Twenty or 30 belt presses, with capacities ranging from 0.1 to 5 t of pulp hour^{-1} have been made. The smallest unit, a scaled-down version of the one described in detail (Davys & Pirie, 1965), is the one most widely used. Its PVC-coated nylon belt is 5 m in circumference, 30 cm wide and passes round two pulleys 36 cm in diameter. One is perforated, and the other is forced outwards by springs able to exert 1 to 2 tf. The belt moves at 2 to 4 m min^{-1}. There is no relative movement between the belt, the pulp, and the perforated pulley. In spite of this advantage, belt presses are not wholly satisfactory for this purpose. Unless the pulley is very large, or run very slowly, pressure is applied so suddenly that the pulp tends to squeeze back along the belt, or out at its sides. Other possibilities are discussed later (p. 146).

The tests made on large-scale equipment in 1940–41 and again in 1948–49 showed that existing screw expellers did not extract juice satisfactorily from unpulped leaves. Nevertheless, the idea of extracting juice in one operation, in a piece of slow-moving equipment is obviously attractive – especially if use is envisaged in a less developed country where an animal might be the source of power. We (Davys & Pirie, 1963) therefore made a batch extractor in which 100 to 300 kg lots of leaf were pulped by a heavy, ribbed roller driven round a horizontal perforated bed. This extracted less protein than the other units, but used less power. Most of the LP used (p. 115) by Singh and his colleagues in the Central Food Technological Research Institute (Mysore) was made there with this machine; it was also used in Nigeria and Papua New Guinea. It has, however, several faults and that precise design would not be copied if simple units were being made now. But the basic principle – that LP will be most useful if it can be made with simple equipment by relatively unskilled people – is sound although it was then premature. In 1958, when the first unit was made, the concept of Intermediate, or Appropriate, Technology had not been formulated and organisations such as the International Bank for Reconstruction and Development and the grant-giving Foundations (e.g. Ford Foundation, 1959) were still obsessed by the idea that the world's food supply could be assured by intensive development of large-scale, sophisticated technology. This outlook did not. change till about 1970. But for that technological obsession, more interest would probably have been taken in our work in the 1960s, and the Rockefeller Foundation grant might have been continued.

The factors that led to continued scientific interest in LP in Britain after 1948, whereas interest diminished in other countries, probably also explain the reawakening of commercial interest. Furthermore, the idea of using LP as food was at that time given some publicity in the popular press. The Angus Milling Co. Ltd (Aberdeen) started commercial production on a small scale. They separated grass into juice and fibre in a standard oil expeller: the dried fibre was 'Graminex Feed' and the coagulum from the juice was 'Graminex'. So far as I know, there is no publication on the process, and only one on the product (Carpenter *et al.*, 1952), but a report written in 1951 by J. L. Dawson showed a comprehension of the principles involved, and the advantages that should result from commercial exploitation, that was unusual for that time. The defects of standard oil expellers were apparent to Powling (1953) and he designed a simplified expeller more suited to the job. This 'Protessor' is still being manufactured and is used in several institutes in Britain and elsewhere. Powling made LP from many different species of leaves, including leaves that are by-products; this was also the approach of Goodall (1950) who used sugar beet tops.

In several institutes (p. 74) unfractionated leaf juice is used as an animal feed. Most of the LP used in human or animal feeding trials was, and still is, made by heat coagulation; the reasons for choosing this method are discussed later (p. 50). The protein in leaf juice can be precipitated, either completely or in fractions, in many other ways, e.g. by acidification or by adding salts, solvents miscible with water and water-immiscible polar solvents. The circumstances in which the use of these methods is advantageous will be discussed later (p. 70). Many leaf extracts coagulate spontaneously on standing for several hours at room temperature. There are few circumstances in which this would be a sensible method for making LP because of the autolytic and oxidative processes that happen at the same time.

3

Choice of crops and yields of extractable protein from them

With a forage, or other leafy crop, the part that is harvested is the part in which protein is synthesized, and not a part to which it is translocated. Because translocation losses are eliminated, a forage crop should, in a suitable climate and given a similar level of husbandry, yield more protein ha^{-1} year^{-1} than any other type of crop. This expectation is borne out in practice. The advantage is even greater if the species used can regrow several times after harvesting so that the ground has a photosynthetically active cover throughout the growing season. That has been agreed for many years and was the basis of Slade's advocacy of LP production. In the late 1930s, when trustworthy amino acid analyses were beginning to appear, it seemed to many scientists (e.g. Chibnall, 1939) that the mixture of proteins in leaves should have good nutritive value. Even without the analyses this was probable because many nonruminant animals, i.e. animals with alimentary tracts in which there is little microbial synthesis of amino acids, live almost exclusively on leaves. Furthermore, the great metabolic and synthetic capacity of leaves forces us to assume that they contain a more extensive array of enzymes than any animal tissue. It seems unlikely that the same amino acid excess, or deficit, will characterise all these different enzyme proteins. The distribution of amino acids in seeds, and other storage organs with few metabolic activities, is more likely to be nutritionally unfavourable. This expectation arises for purely statistical reasons, it does not depend on the invalid assumption, which has crept into some articles, that enzyme proteins are in some way 'better' than storage proteins – there are simply more different types of them in an active organ so that an excess or deficit of any one amino acid is unlikely.

Though not always explicitly stated, the two cardinal proposi-

15

tions, that LP is potentially the most abundant source of protein, and that it should have good nutritive value, stimulated all the research outlined in the preceding chapters. Those who doubted the value of research on LP, or who were actively hostile to it, seem always to have paid too little attention to the basic propositions: they did not contest them, they simply disregarded them. Undoubtedly, difficulties could be foreseen. Otherwise LP production would not have been a research project. It may not be possible to maintain a useful green cover on the land in a practical farming system. It may not be possible to extract protein with an economically realistic expenditure on machinery and energy. Other components of leaves may combine with, or damage, the protein in the course of extraction so that some essential amino acids are destroyed or made unavailable to nonruminants. And so on. These are matters on which research is still needed.

The technique of extraction

Once the feasibility of extraction had been demonstrated, attention could be turned to selecting suitable leaves. The extractability of LP obviously depends in part on the vigour with which extraction is undertaken and the criteria used to distinguish extracted protein from protein still carried by the leaf fibre. When, as is usual in the laboratory, pulp is squeezed through a piece of cloth, the thickness and character of the weave are relevant, and also the amount of water added during pulping. Similarly, when pulp is centrifuged, results depend on intensity and duration. Prolonged centrifuging sediments chloroplasts and their fragments; both should be included in the category 'extracted LP'. With sufficiently prolonged grinding, e.g. in a motor-driven, end-runner mill, the leaf fibre can be so thoroughly disintegrated that it all passes into the supposed LP fraction. It is likely that nearly complete distintegration of the fibre explains occasional claims that nearly all the protein can be extracted, without enzyme digestion, in the laboratory. On the other hand, as already mentioned (p. 7), some methods of milling can attach protein to the fibre. For all these reasons, some standardisation of extraction technique is essential, and the techniques used should have at least a qualitative resemblance to techniques that would be feasible in actual large-scale extraction.

Early routine measurements of extractability were usually made by pulping in a domestic meat mincer, squeezing out juice from a weighed amount of pulp by hand through cotton cloth of the type used for pocket handkerchiefs, remincing the fibre with added water, and squeezing again. To minimize sampling errors, the water and N contents of the original leaf were measured on a sample of the first pulp. After measuring the volume of the mixed extracts, the amount of protein N and nonprotein N in it were measured by adding an equal volume of 10% trichloroacetic acid (TCA) to a sample, centrifuging, and analysing the precipitate and supernatant. When working with leaves from species containing large amounts of nonprotein N, the precipitate should be resuspended in dilute TCA and centrifuged again. The extracted fibre was dried, weighed, ground, and analysed for N. From these measurements, the percentage of the total N appearing in each of the three fractions can be calculated, and the precision with which the operations were carried out can be assessed from the agreement between the mass of N in the sample of leaf taken and the sum of the masses of N in the three fractions. More N is precipitated from leaf extracts by TCA than by heating. The difference is usually 5 to 10% but it can reach 20% with extracts from very young leaves. The more rapidly the extract is heated, the more nearly the N precipitated by heating approaches the amount precipitated by TCA; the longer the interval between making the extract and adding TCA, the more nearly the amount precipitated by TCA approaches the amount precipitated by heating. These differences mainly arise because TCA precipitates the nucleic acid in ribosomes, and nucleic acid is destroyed by leaf ribonuclease during ageing or slow heating (Pirie, 1950; Singh, 1960).

Although much experience, and background information, was gained from measurements made by that method, domestic mincers have many defects. The individual plants in many crops are so large that at least 3 kg should be taken as a representative sample from a plot. Pushing that amount of material through a mincer is tedious and, with fibrous leaves, frequent stops may be necessary to disentangle fibre from the cutter. The juice liberated from lush leaves fills the barrel of the mincer and may, unless great care is taken, leak out backwards instead of coming out along with the pulp. It is difficult to make hand squeezing quantitative and consistent. Operators vary in persistence and in the extent to which they rearrange the charge within a cloth while squeezing. With financial support from the UK committee of the IBP, a pulper and press were made which gave more systematic results (Davys & Pirie, 1969; Davys *et al.*, 1969).

Research on the same subject and using the same methods, but in different climates, was one prime objective of the IBP. A study of the attainable annual yields of LP seemed a peculiarly suitable theme. Results are more likely to be consistent, and comparable although coming from different institutes, if pulping is made extremely thorough. The old (Pirie, 1961) method of forcing leaves through a narrow slot could have been used – but only with difficulty on a sample large enough to be representative of a crop. We therefore chose a pulper with beaters running at a tip speed of 58 m s⁻¹. It probably liberates protein from the leaf structure more completely than would be feasible in commercial production; but it sets a standard.

Careless operators, when grinding dry samples in a hammer mill, may leave some material in the mill at the end of a run and this will usually not have the same composition as the more easily disintegrated material that passes through the mill. This type of error must be guarded against in designing an analytical pulper. The possible error becomes smaller the larger the ratio of the mass of material pulped to the mass remaining in the pulper. At the end of a run, the IBP pulper contains 200–300 g of material. Experiments in which pieces of paper were put in from time to time along with the leaf showed that after 500 g had come through, equilibrium was established between emerging and retained pulp. The first 500 g of pulp should therefore be discarded.

The IBP pulper is a stepped drum 44 cm long, 27 cm in diameter at the feed end and 32 cm at the discharge end; the difference in diameter ensures an air flow in the direction of movement of the pulp. A rotor inside the drum carries 58 fixed beaters with a 2-mm clearance from the drum. In the laboratory, the rotor is driven by a 5-HP motor; the speed and direction of rotation are so arranged that the pulper can also be mounted on a 'Landrover' and driven from the power-take-off. It can therefore be used in the field. For safety the pulper is fed through a 5.5-cm diameter tube. Helped by a plunger so shaped that it cannot reach the rotor, feeding at up to 1.5 kg min⁻¹ is possible. When used by operators who can be trusted to work safely, a wider entry tube can be fitted, thus permitting faster feeding. The outer drum is easily taken off so that all parts of the machine in contact with the crop can be cleaned.

LP for most of the early human feeding trials (p. 117) was made from pulp produced by the IBP pulper, pressed in the small belt press already described (p. 13). That press is unsuitable for quantitative work. Pressed fibre from it is about 5 mm thick and it is exposed to 1.5 to 2.0 kgf cm^{-2}. the IBP press simulates these conditions. Because pulp is homogeneous, there is no need to work with a large sample. Because conditions at the edge of material being pressed differ from those in the body of the cake, the larger the area of cake the better. In the IBP press 900 g of pulp is spread evenly on a cloth on a square, grooved platen 23 cm each way, the cloth is folded so as to make the area of pulp 450 cm^2 and another grooved platen is placed on top. The assembly is mounted with the grooves vertical and subjected to a pressure of 1 tf applied by means of a bell-crank with a 40 to 1 ratio. The dimensions were chosen so that the unit can be carried and, like the IBP pulper, can be put in a 'Landrover' and used in the field (Davys *et al.*, 1969).

The hard cake of fibre that is made by the IBP press cannot be quickly and easily broken up for re-extraction. There is therefore no second extract to mix with the first, but mixing is as important as with hand squeezing because the composition of juice coming out during the first few seconds differs from that coming out during the 2 to 3 min for which pressure is maintained. Pulp, fibre and extract are analysed as already described. Because pulping is more thorough in the IBP pulper than in a mincer, more protein is released from cells – especially with rather dry or fibrous leaves. On the other hand, because second extracts are not made, less of the released protein appears in the juice, and a correspondingly larger amount of the leaf N is retained in the fibre. From the soft, moist leaves usually studied, the IBP system extracts 20 to 30% less protein than the older system. This discrepancy is not necessarily a defect; neither system reproduces faithfully the conditions of actual large-scale operation. Measurements made with these agronomic tools are intended to arrange crops and systems of husbandry in the order of their probable productivity; they cannot give precise information about the performance of large-scale equipment.

A crop harvested in a manner that does not bruise the leaves, does not deteriorate in a few hours in a cool climate. Deterioration is rapid after harvesting with a flail or 'precision chop'. It gets faster as the temperature increases. Batra *et al.* (1976) measured the extractability of LP from seven crops kept at 28 to 35 °C for various times after harvesting. The yield from most of them was nearly halved after 9 hours. They attributed most of the diminution to the evaporation of water from the crop so that there was insufficient juice to carry LP out of the pulp (p. 29). However, sprinkling the crop with water or dilute alkali (Jadhav & Joshi, 1982) did not significantly protect the LP. Until more is known about the cause and extent of deterioration, it would be prudent to harvest no more plots in an experiment than can be processed within an hour. For similar reasons, the condition of a crop early in the morning and late in the afternoon is not the same – especially in a hot, dry climate. Judgement is therefore necessary in agronomic work and plots should be replicated so that similar ones can be taken at different times of day and be pulped after different amounts of delay. After pulping, speed is essential because protein immediately begins to coagulate onto the fibre and to autolyse in the extract. A delay of 2 hours at 24 °C between pulping and pressing diminished the yield of protein from red clover (*Trifolium pratense*) to half (Davys *et al.*, 1969); Tracey (1948) and Singh (1962) got a similar diminution with wheat (*Triticum aestivum*). Pulp should therefore be pressed within minutes of being made.

Because of autolysis, there should be no delay in quantitative work during the operations of measuring the volume of juice coming from the 900 g of pulp in the IBP press, mixing it, taking a sample and precipitating it with an equal volume of 10% TCA. Thereafter delay is immaterial. A crop, or potentially harvestable wild growth, can therefore be pulped, pressed and sampled with the IBP equipment at any site accessible to a 'Landrover' and analyses can be completed at leisure in the laboratory.

Measurements made with the IBP pulper and press are repeatable because the pressure applied is always the same and it is applied at approximately the same rate. The second point is important because chloroplasts and their fragments are held back by tightly packed fibre. Juice made by sudden application of pressure, especially if, as in most hydraulic presses, the layer of pulp that is being pressed is many centimetres thick, will be depleted of protein. The amount of protein extracted depends on the water content of the leaf, i.e. on the amount of fluid available to carry liberated protein away from the mass of fibre. The IBP procedure measures the *expressible* protein (Butler, 1982). To measure the potentially *extractable* protein, pulps should be diluted to a standard water content, say 90 or 95%, and samples of juice should be separated from them by simply pouring on to a cloth without applying any pressure. Although that procedure exaggerates the amount of LP that could be extracted in practice, it gives a more physiologically meaningful result. The merits and defects of equipment which is, or could be, used to extract leaf juice on a commercial scale are discussed later (p. 146).

Selection of species, and intraspecific variability

The tables that have been published listing the extractability of protein from leaves of many species can be very misleading because extractability depends on many factors besides species. Early experience (e.g. Crook, 1946; Crook & Holden, 1948) justified the tentative generalisation that the greater the percentage of water in a leaf, and the greater the percentage of protein in its dry matter (DM), the greater the percentage of that protein that will be extractable from a given species by a given technique. This generalisation still seems to hold. A striking example was the observation (Arkcoll & Festenstein, 1971) that the yield of LP m^{-2} was smaller from a plot of kale (*Brassica oleracea*) given phosphorus and potassium than from an unfertilised plot because, although the DM yield was increased 68%, the percentage of N in the DM was diminished by 56%. Obviously, the generalisation holds only for comparisons between different treatments given to plots of the same species; species containing more protein do not necessarily extract better than others. It does however follow that there is a disproportionate advantage in harvesting leaves that are young, well manured with N, and well watered. They will not only contain more protein, but more of that protein will be extractable. However, the yield from each harvest is smaller. For example, with four varieties of lucerne the LP contained 55% of true protein when made before budding, and 47% after flowering, but the yield was twice as great after flowering (Bubicz & Jelinowska, 1983). Nazir & Shah (1985) recorded a steady diminution in total yield from Persian clover (*Trifolium resupinatum*) as the interval between harvests was lengthened from 20 days to 100 days.

Species differ in the extent to which maturity diminishes extractability; there are also differences between varieties. Crook & Holden (1948) could extract protein from one variety of strawberry (*Fragaria vesca*) but not from another. Byers & Sturrock (1965) compared five maize (*Zea mays*) varieties and found that the differences were not the same in successive years. Wheat varieties differ (Arkcoll & Festenstein, 1971). Differences in the date at which the haulm of different potato (*Solanum tuberosum*) varieties dies are well known and they affect the amount of protein that can be extracted from haulm if that is taken as a by-product when tubers are lifted (Carruthers & Pirie, 1975). Fat hen (*Chenopodium album*) shows extreme variability. Arkcoll (1971) singled it out as a species which continued to give good protein extraction after flower development; the variety used by Heath (1977) matured more abruptly than three other species with which it was compared; Carlsson (1975) found one variety yielding more than three times as much m^{-2} as another, and 80% of the protein was extractable from an Austrian variety, compared to 50% from a Swedish variety. The yields from 15 varieties of fenugreek (*Trigonella foenum-graecum*) (Chandramani *et al.*, 1975*b*) and five of radish (*Raphanus sativus*) (Bagchi & Matai, 1978) differed by a factor of five. These radish leaves were taken at times of normal harvest: there was little correlation between root and LP yields. Leaves from cassava (*Manihot esculenta*) varieties, between which Fafunso & Oke (1977) found differences, were not taken at the time of normal harvest. Hegsted & Linkswiler (1980) found more saponin in LP from saponin-rich varieties of lucerne, and the LP was less acceptable to rats. Yields from six lucerne varieties differed little (Jelinowska & Magnuszewska, 1981), similarly with the four varieties studied by Kehr *et al.* (1979). Lundborg (1980*a, b*) found differences between 11 varieties of sunflower (*Helianthus annuus*) greater than those between *Brassica oleracea* varieties. In spite of ensuring that leaves of the same age, and taken from similar positions on the plant, were compared, De Jong (1984) found considerable differences among 17 tobacco (*Nicotiana tabacum*) varieties in both protein extractability and in the properties of the extracted protein.

The ratio of protein N to nonprotein N differs between species; it also varies within a species with the conditions of husbandry, atmospheric temperature and possibly the time of day at which the leaf is harvested (Chibnall, 1939; Ostrowski *et al.* in Wallace, 1975). Because nonprotein N will be more difficult to use profitably in commercial practice (p. 132) than N in LP or associated with the fibre residue, species and conditions should be chosen in which nonprotein N is as small a fraction of the total N as possible.

Species that appear to extract well in laboratory conditions may be almost impossible to handle on a large scale, either because the extract is glutinous (e.g. comfrey *Symphytum asperrimum* and *Symphytum officinale*) or because it forms an intractable mass of froth. Some varieties of lucerne are difficult to handle for the latter reason. After working for a few hours the equipment may be difficult to find and almost impossible to manipulate. By contrast, some of the grasses that extracted badly when put through domestic mincers extracted well in the more violent conditions of the IBP pulper.

It follows from all this that a satisfactory basis for commercial LP extraction will not be established until there has been more agronomic, or plant physiological, research. Sowing dates, fertiliser treatments, time of harvest and ability to regrow after harvest will have to be studied with more species and varieties. There is also scope for plant breeding. The crop plant varieties used so far were all selected according to criteria, such as the ability to yield abundant seed, that may well be inimical to LP yield. The varieties most useful for LP production will probably have flowering delayed or prevented by sowing them at unusual times, or in unusual latitudes, or by genetic manipulation, or through use of growth regulators, so that senescence is delayed and there is a prolonged period of vegetative growth. No attempt has so far been made by plant breeders to use LP yield as a criterion in selecting candidates for research on breeding.

When species and varieties are being selected for detailed study, the points already made should be borne in mind, i.e. there should be abundant, lush, protein-rich leaf. There are other favourable qualities. The leaves should not be carried on a very fibrous stalk or the energy needed for pulping will be excessive unless some form of scutching is possible. The leaf should be neutral or slightly alkaline; although acidity can be partly counteracted by pulping with added alkali, this complicates the process. The presence of tannins and phenolic substances diminishes protein extraction; there is enough in some leaves to prevent protein extraction completely (Bawden & Kleczkowski, 1945). Butler (1982), in a paper which makes the useful distinction between LP *expressibility*, i.e. the amount which comes out in the juice present in the leaf, and *extractability*, i.e. the amount which comes out when a suitable volume of extra water is added, lists phenolic content as the factor most often responsible for small yields of LP.

So as to give an impression of the potentialities of LP production, it is worthwhile surveying some results that have been published on protein extractability and yield although little information is often given about the antecedents of the plants used, and the most suitable varieties may not have been chosen. It is convenient to group possible sources of leaf arbitrarily into the categories:

Species used in conventional agriculture.

Leaves available as the by-product of a conventional crop.

Tree leaves.

Water weeds.

Miscellaneous unconventional species of leaf.

Conventional species

During the period in which extraction equipment was being designed and improved at Rothamsted, crops grown in the normal manner were used; latterly many experiments were on cereals sown two to four times more thickly than usual and fertiliser was used on some plots unusually liberally. This departure from normal farm practice was permissible because crops were harvested from late April until early June while still green and without a heavy head; there was therefore no risk of lodging. Our experimental plots were never irrigated although in this part of Britain there is a summer water deficit four years out of five. The results were collected in two papers (Byers & Sturrock, 1965; Arkcoll & Festenstein, 1971). So as to make use of winter sun and ensure an early start in spring, much of this work was on autumn-sown cereals. Proper timing of the first harvest is important because it affects the amount of regrowth. Species differ in the vigour of regrowth: with wheat it is excellent, with barley (*Hordeum vulgare*) less, with rye (*Secale cereale*) poor, and with maize negligible. When two, or sometimes three, harvests of short, leafy material were taken, the yield of LP was larger than the maximum yield obtainable from a single harvest; obviously, the amount of labour expended in harvesting was also greater. The land was then ploughed and sown with a variety of summer crops. The results of one such experiment are given in Table 1. More than 2 t of LP, i.e. 334 kg of protein N, was produced from the same piece of land within a year. Figure 1 shows the yields in successive years in experiments of this type. The increase in yield was the result both of increasing skill in harvesting crops at the correct time, and of improvements in the extraction equipment. The figure also shows the dependence of large yields on adequate summer rainfall.

Grass did not extract satisfactorily in early forms of equipment. With improved equipment it became an excellent crop and had the advantage of not necessitating repeated sowing. The well-known beneficial effect of frequent cutting was shown clearly. Figure 2

Table 1. *Yields of extracted protein (LP) in kg ha^{-1} from Arkcoll & Festenstein, 1971)*

Winter wheat sown 20 Oct. 1967,		harvested 1 May 1968	819
		regrowth 18 June 1968	284
Land then ploughed and sown with either:			
Mustard (*Sinapis alba*)		Fodder radish	
First crop harvested		First crop harvested	
1 Aug. 1968	426	12 Aug. 1968	460
Second crop harvested		Second crop harvested	
29 Sept. 1968	490	15 Oct. 1968	461
Total yield	2019	Total yield	2024

Figure 1. The yield of extracted protein (nitrogen × 6) from a succession of crops grown on the same plot without irrigation. Improvements to the extraction equipment, and increased skill in harvesting at the right time, contributed about equally to the increased yield between 1960 and 1968.

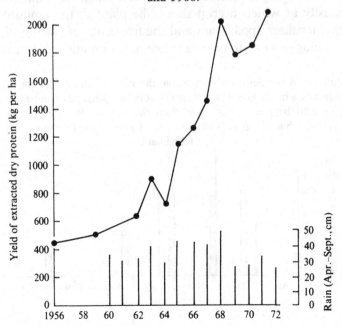

shows one such experiment on cocksfoot (*Dactylis glomerata*) harvested from different plots at three different ages, compared with five harvests from the same plot. The summed yield from five harvests was three times the maximum from a single harvest, and 1969, as Figure 1 shows, was a dry year. Figure 3 illustrates the general phenomenon of the decline in yield of extractable protein at about the time of flowering. It has been commented on in many papers from many different laboratories but, as already pointed out (p. 22), there are exceptions.

In all these experiments, increasing the amount of fertiliser N up to 264 kg ha^{-1} increased the yield of LP, although there was little increase in the yield of DM after about 132 kg ha^{-1}. Very large doses of N are, however, used inefficiently. Thus the largest increase in the yield of LP on going from 132 to 264 kg N was only 76 kg, i.e. only about 10% of the additional N was recovered in LP. It is likely that it would not, as a rule, be economic to aim at these very large yields. The maximum yields from legumes depending for N on their root nodules, and without irrigation, were 1247 kg ha^{-1} with red clover, and 1009 with lucerne, each harvested three times.

Heath (1977) and Heath & King (1977) pointed out that in cocksfoot, ryegrass (*Lolium perenne*) and fodder radish, true protein accumulates mainly in the laminae. Accumulation, therefore, diminishes, or stops, when stem growth replaces laminar growth. The density at which a crop should be planted, the amount and timing of fertiliser application and the frequency of cutting, should all be managed so as to ensure maximum formation of leaf rather

Figure 2. A comparison showing that the yield of extractable protein from single harvests in 1969 of cocksfoot grass taken at different dates (the solid bar) was much less than the sum of the yields from successive harvests (total yield was 1670 kg ha^{-1}) from the same plot (hatched bar).

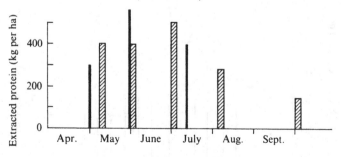

than stem. Inclusion of large amounts of stem in material from which LP is being extracted contributes little extra protein, and it acts as an absorbent mass within which the laminar protein gets lost. Heath contrasted the extra trouble and expense of repeatedly sowing annual crops, with the apparent economy of using perennial crops; he concluded that annuals give more flexibility, can probably be managed so as to give a more even supply of material throughout the season, and should in the conditions of commercial farming yield 840 kg of LP ha^{-1}. This is more protein than any other system of farming produces, and it does not take into account the value of the fibre residue and the nonprotein N in the juice.

With irrigation, but without N fertiliser, lucerne in New Zealand

Figure 3. An indication of the manner in which the extractability of protein declines as leaves age, so that there is a definite optimum date for harvesting. Downward arrow indicates time of flowering of crop.

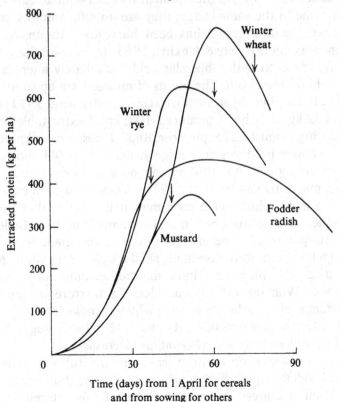

Time (days) from 1 April for cereals
and from sowing for others

yielded 2 t of LP during seven summer months when processed in the IBP pulper. There was little difference in LP yield between four-week and five-week harvests; the latter gave larger yields of total DM (Allison & Vartha, 1973). This is an important point when the residue is thought of as a valuable fodder (p. 129). During the New Zealand winter, ryegrass (*Lolium multiflorum*) yielded 1.2 t of LP ha^{-1} from six harvests in 4.5 months. As at Rothamsted, the extra yield from more than 180 kg of fertiliser N ha^{-1} was gained wastefully (Vartha & Allison, 1973). From the standard mixture of pasture species used in northern Victoria (Australia), and given irrigation and a top dressing, Ostrowski-Meissner (1983a) extracted 1.4 to 1.7 t of protein annually.

Other temperate zone conventional crops have been less thoroughly studied. Byers & Sturrock (1965) compared five clover (*Trifolium* and *Melilotus*) varieties. Apart from their ability to fix N, they seemed no better than the other species that were being used at the time; 40 to 50% of their protein was extractable. Tares (*Vicia sativa*) came in the same range; they are a useful summer crop to sow after a winter cereal has been harvested. An unsuccessful attempt was made (Byers & Jenkins, 1961) to increase the yield by spraying the crop with gibberellic acid immediately after the first harvest had been taken. The merits of mustard are apparent from Table 1. In another experiment (Arkcoll & Festenstein, 1971) a plot given 132 kg of N ha^{-1} produced 10.6 kg of extractable protein daily during August and September. Rape (*Brassica napus*) produced less. Sunflower produces very large yields of both DM and protein if given enough N, but this protein does not extract well from mature plants (Lexander *et al.*, 1970). These measurements were made on leaves that had been frozen. It is possible that freezing coagulated some of the protein and so made it unextractable, but earlier (unpublished) experiments at Rothamsted, made on material pulped while fresh, were essentially similar unless very young plants were used. We may here have another example of a varietal difference: Williams (1979) considered a mixture of sunflower and fodder radish the best crop from which to make LP. On the basis of laboratory scale extractions, Rawate & Hill (1985) suggested the use of Jerusalem artichoke (*Helianthus tuberosus*).

There is research on LP in at least nine institutes in India, and so much reliable quantitative agronomic information comes from there, that a digression on the origins of this interest may be relevant. In 1959, in casual conversation, I had criticised some aspects of the Bhoodan Movement in India and commented on the place of technology in a rationally run society. My opinions were quoted to Jayaprakash Narayan, who happened to be in London at the time. He asked me to come and see him, and showed so much interest in research on LP that I sent him some papers, which he gave to Jawaharlal Nehru, who asked the Council for Scientific and Industrial Research to look into the matter. I was invited to India at the beginning of 1961; research started soon after that in the National Botanic Gardens in Lucknow and in the Agricultural College (now University) in Coimbatore. Soon after the start of the IBP, research in Coimbatore was complemented by research in Aurangabad and Calcutta, using IBP equipment donated by the UK committee of the IBP (Pirie, 1976*b*).

Lucerne, given irrigation water and some N fertiliser, yielded more than 3 t of LP ha^{-1} in a year at Aurangabad in spite of attack by aphids (Dev *et al*, 1974). More than half was from six to eight harvests taken in the six months October–March. Several additional points were made in that series of experiments: the yield was increased by sowing in rows that were 23 rather than 46 cm apart. The yield of LP was increased 47% in the first year, and 60% in the second, when lucerne was irrigated during summer (Savangikar & Joshi, 1976). In some experiments, growth regulators such as 'Simazine' increased the protein content of leaves: 'Simazine' did not increase the yield of LP from lucerne (Dev *et al*., 1974) or hybrid napier grass (*Pennisetum typhoideum* × *P. purpurea*) (Gore & Joshi, 1976*b*). Earlier work by the same group (Joshi, 1971) had shown that inoculating lucerne seed with rhizobia increased the yield of LP by 12 to 19%. In some parts of India, lucerne does not survive well; it is treated as an annual rather than a perennial crop. Until the reasons for this poor survival have been found, and possibly corrected, lucerne lacks the main advantage it has elsewhere. In a survey of 1000 trials made during 12 years, Joshi & Mungikar (1983) concluded that, of the nine variables studied, the water content of the crop at the time of harvest had most effect in promoting LP extraction. Yields from other crops which occupy ground for a whole year or, outside the tropics and sub-tropics, for the growing season, have been less thoroughly studied. There seems to be general agreement (e.g. Deshmukh *et al*., 1984) that tropical grasses such as *Sorghum* do not extract well enough to be probable sources of LP unless they are harvested when very young. Thus, annual yields from hybrid Napier grass varied from 2.2 to 0.6 t ha^{-1} in spite of liberal use of fertiliser N (Gore *et al.*, 1974). The variation probably depends on the precise interval between successive harvests; this crop quickly becomes fibrous which, as already explained (p. 23), wastes energy during extraction and diminishes extractability. The yield of DM is greatest with 50-day intervals between harvests, the yield of LP is greatest with 18- to 23-day intervals. The largest recorded annual yield of LP was 0.98 t ha^{-1}, but reasons are given for thinking that 2 t would be attainable. Yields in West Bengal (Matai, *et al.*, 1976) were consistently smaller than those in Aurangabad. Yields from the three grasses *Bracharia mutica, Cenchrus glaucus* and *Panicum maximum*, (pearl millet) grown further south in Coimbatore (Balasundaram *et al.*, 1975; Chandramani *et al.*, 1975a), depended on the frequency of cutting, *P. maximum* gave 1.3 t of LP ha^{-1} in a year when cut every 30 days, with a daily yield of 15.8 kg during one month, whereas the yield was only 0.5 t when the interval between harvests extended to 45 days. Millet in West Bengal behaved similarly (Matai *et al.*, 1980).

Measurements on perennial crops, or on a succession of crops grown on the same piece of land, justify the claim that, with reasonable skill, LP extraction gives an unprecedently large annual protein yield. Experiments which last for only part of a year give useful information about the possible suitability of a species. Carlsson (1983) and Telek & Martin (1983) list more than 100 such experiments. These lists need not be repeated here.

Although the efficacy of green manuring is doubtful in regions with high soil temperatures, the technique is widely used. Species are chosen which give an abundant yield of N-rich foliage quickly: that is the basic quality of a good LP source. If green manuring is thought desirable, it seems better to plough in only the two by-products of LP extraction. Some of the crops already discussed, e.g. lucerne and mustard, are used; so are a few others with suitably soft leaves, e.g. kudzu (*Pueraria phaseoloides*) (Byers, 1961) and *Sesbania sesban* (Gore & Joshi, 1976a; Jadhav *et al.*, 1979). The latter extracted well, but was not thought preferable to lucerne or cowpea (*Vigna sinensis* or *unguiculata*). However, the extensive range of tropical legumes commonly used as cover crops, green manures, or hedge plants has not yet been adequately studied. This is a promising group of species and the techniques for propagation are already well known.

Byers (1961) noted good extraction of LP from cowpea. It has many merits: even without washing, its LP has a pleasanter flavour than LP from lucerne, and its associated rhizobia do not waste 40 to 60% of their metabolic energy, as those associated with soya and lucerne do, in making hydrogen rather than fixing N (Schubert & Evans, 1976). In spite of good response to inoculation (Deshmukh & Joshi, 1973), it was usually given both farmyard manure and 20 kg N ha^{-1}. More N gave no extra yield. In summer, with that treatment, it yielded 895 kg LP ha^{-1} from three harvests during 80 days, i.e. it gave 11.2 kg of LP daily ha^{-1} (Deshmukh *et al.*, 1974). An earlier unpublished experiment gave a daily LP yield of 16.9 kg ha^{-1}. If that could be repeated it would be a world record! Such yields should be borne in mind before paying attention to the ludicrous daily yield of 0.5 kg ha^{-1} (Anonymous, 1965) in a gloomy forecast of the potentialities of cowpea as a source of LP in Jamaica.

Tetrakali, the tetraploid form of *Phaseolus aureus*, grows well in the wet season in West Bengal where it is used as fodder. It produces LP at rates which correspond to 1760 kg ha^{-1} if growth could be maintained by irrigation during the dry season. Bagchi & Matai (1976) concluded that it is an excellent crop which, on better soil than they have access to, would give still larger yields. Berseem (*Trifolium alexandrinum*) is a familiar forage crop in Egypt, India and Pakistan. Its use as an LP source is being studied in many institutes (references in Telek & Graham, 1983). It gives yields similar to those of cowpea, lucerne or tetrakali if it is harvested at 20-day intervals, but protein extractability declines rapidly as the crop matures (Mungikar *et al.*, 1976a; Shah *et al.*, 1976). Its soft leaf makes pulping economical.

The merits of three chenopodiums, *Atriplex hortensis, Chenopodium quinoa* and *Amaranthus caudatus* are detailed by Carlsson (1980), and those of four brassicas, *B. campestris, napus, juncea* and *carinata*, by Lyon *et al.* (1983). After 3 or 4 months growth, the yield of extracted protein could reach 800 kg ha^{-1}.

By-product leaves

The idea of using as a source of LP, leaves that are the by-product of some conventional crop has an obvious appeal. Unfortunately, the range of potential sources is limited. When a dry seed such as maize, rice (*Oryza sativa*) or wheat is harvested, the leaf is withered. Sugar cane (*Saccharum officinarum*) is a possible source. The tops are now usually burnt because, by removing trash, this makes hand-cutting easier, and by removing snakes and scorpions, pleasanter. With mechanised cutting these advantages of burning lose some cogency; furthermore, burnt cane deteriorates rapidly and must therefore be harvested and processed quickly or there will be loss of sugar. However, even in regions that are so humid that the tops are still green and fairly moist at the time of harvest, they seldom contain more than 1.2 % N (in the DM). That, however, corresponds to at least 0.5 t of protein ha^{-1}; Balasundaram *et al.* (1974*a*) extracted 108 kg of LP ha^{-1}. Li Sui Fong (1982) and Deepchand (1984), in Mauritius were able to extract up to 8 kg from 1 t of cane top by intense pulping with added water. However, cane tops are so fibrous that the energy needed for extraction may be dispropor-tionately large. If the tops were used as a fibrous material with which to mix leaves such as those from sweet potato or water hyacinth, which are so soft that they are not easily handled in existing machines, some protein would be extracted from the cane tops as well.

The area already devoted to cassava (tapioca, manioc, or yuca) is so large, and there are such extensive plans for increasing production both for food and for industrial alcohol, that thorough investigation of its leaf would be worthwhile. Although there are 35 page references to cassava in the book edited by Telek & Graham (1983), there has been little advance on the early work of Byers (1961), Singh (1964) and Balasundaram *et al.* (1974*a, b*). The last of these papers puts cassava into 'category 1' in a classification of 212 species, but does not state clearly that the leaves used were taken at the time of normal harvest of the tubers. Efficient extraction equipment was not used in these experiments: that is an important point because cassava leaves when mature are rather dry and tough. Other by-product leaves, from which it is not likely that protein could be economically extracted, are those cultivated for the sake of essential oils or perfumes; they are usually subjected to drying or steam distillation, and these processes will coagulate the protein *in situ*.

In the temperate zone the two most abundant leafy by-products are sugar beet (*Beta vulgaris*) and potato. The former is largely wasted in Britain (about 80% of it is usually ploughed in) but not in Germany or Poland. The latter is wasted in almost all countries. An argument sometimes advanced for ploughing-in these by-product leaves is that they have manurial value. That is partly true. But

Widdowson (1974) showed that the extra yield of barley in the following year attributable to ploughing in 130 kg of N in sugar beet tops, would have been given by 8 kg of fertiliser N. Goodall (1950) made a product containing 38 to 40% protein on a commercial scale. At Rothamsted, LP containing 60% protein was made (Pirie, 1958). The maximum yield quoted was 500 kg ha^{-1} but the yield obviously depends on weather and on the date of harvest. However, even if the yield in practice were only half our experimental yield, the 193000 ha on which sugar beet is grown in Britain would Produce 48000 t of extracted protein as well as a fibre residue that would be more attractive as cattle fodder than untreated tops. According to one estimate, sugar beet tops in the EEC contain 13 Mt of DM, i.e. > 2 Mt of protein, of which half should be extractable.

The outstanding photosynthetic efficiency of sugar beet is in part a consequence of the vertical disposition of its leaves. Towards the end of the growing season, the leaves bend over so that there is more mutual shading and a decline in efficiency. Some experiments suggest that part of the leaf can be harvested early in the year with little loss of sugar – presumably because the new growth remains more nearly vertical. Other examples of partial leaf harvest without loss of the conventional product will be discussed later (p. 37). Before LP from sugar beet can be used as human food, research is needed to find out whether the usual technique of preparation (e.g. Morrison & Pirie, 1961) removes oxalate.

As a safeguard against blight, and to facilitate tuber-lifting later, potato haulm is usually destroyed mechanically or with a herbicide at the beginning of September. It would not be difficult to design a haulm-harvester to work on potatoes ridged up in the usual way; it would be more convenient to harvest haulm from a level field, and there is some reason to think that the main merit of ridging is simply the suppression of weeds. If so, weeds could be controlled with herbicides, and the haulm could be harvested like an ordinary forage crop one or two weeks before the tubers were to be lifted. The main reason why potato haulm is disregarded in Britain as a forage crop is fear that stock will be poisoned by solanin and other glycoalkaloids. These are separated in the 'whey' when LP is made (pp. 60, 104).

The yield of LP depends on the potato variety and the data on which haulm is taken (Carruthers & Pirie, 1975). Some early varieties yield 600 kg ha^{-1}, maincrop can yield 300 in late August, but yield diminishes to 100 or even less by mid-September. In

Poland, Hanczakowski & Makuch (1980) got yields of 31 to 155 kg ha^{-1} from ten varieties; in India Srivastava *et al.* (1984) studied eight varieties. The largest yield was 371 kg ha^{-1} (149 kg of actual protein) after 80 days. They, and Hutchinson (1978), found that harvesting at that time diminishes tuber yield by about 200 kg ha^{-1}. It would therefore seldom be worthwhile harvesting early purely for the sake of the extra LP. However, once the idea is accepted that there is something valuable in haulm, prudent farmers will probably take haulm a little earlier as an added precaution against blight. A trend towards earlier lifting is already foreseen in Britain (Ivins, 1973) and is already the practice in Australia (Sale, 1973). In Britain, early potatoes occupy 30 000 ha and maincrop 230000. It is reasonable to conclude that about 50000 t of LP could be extracted from the haulm. Soon after potatoes were introduced into Britain, Sir Thomas Overbury is said to have remarked: 'The man who has not anything to boast of but his illustrious ancestors is like a potato, the only good belonging to him is underground'. The sentiment is admirable, but we may be taking the simile too literally.

At one time, peas (*Pisum sativum*) for canning and freezing were harvested on the vine and the whole mass was taken to a factory, the peas were separated and the vines carted back to the farm. Mobile viners now go into the field and drop the vines as they go along. The old arrangement was well adapted to protein extraction. From pea haulm grown on an experimental plot we got 600 kg of LP ha^{-1}; this is more protein than is present in the peas. Pea haulm collected from a factory 30 km from Rothamsted did not extract so well as fresh haulm – perhaps because there was an interval of about three hours before the bruised and battered haulm was pulped. The percentage of haulm N that was recovered as protein N varied from 22 to 47 (Byers & Sturrock, 1965). More recent research in Spain and Venezuela on extracting LP from pea haulm is reported in the book edited by N. Singh (1984). To make pea haulm a useful source it will probably be necessary to revert to the old method of vining in a factory rather than in the field. Protein could then be extracted without delay. However, peas may not ultimately be a useful source of LP: there is little leaf on some new varieties.

When maize is allowed to ripen completely, the leaves are too dry and depleted of protein for satisfactory extraction. From some varieties, harvested at the end of August, only 20% of the protein was extractable (Byers & Sturrock, 1965). When harvested earlier, i.e. at the sweet-corn stage, nearly half the protein was extractable and the yield of extracted protein reached 480 kg ha^{-1}. The area devoted to sweet-corn in Britain is too small for it to be an important source of LP there.

Vegetable discards in the USA in 1948 had a wet weight of 4 Mt (Willaman & Eskew, 1948); the weight in 1958 was 21 Mt (Oelshlegel *et al.*, 1969), containing an estimated 0.4 Mt of protein. There have been no similar estimates recently in Britain: in 1951 the wet weight was 0.5 Mt, the weight sometimes quoted now is 4.6 Mt (Raymond, 1977). The area devoted to open-air vegetables in England and Wales is 170 000 ha, i.e. intermediate between sugar beet and potatoes. The brassicas alone (Brussels sprouts, cabbage, calabrese, cauliflower, sprouting broccoli, etc.) occupy 60 000 ha. The weight of discarded material is not recorded but is probably more than 200 000 t containing more than 3000 t of protein. Material discarded in the field would be fresh and worth extracting; discards at the retail level may possibly not be worth collecting. Much of the leafy waste in the field is already being collected and some market gardeners pay to have it disposed of.

Tekale & Joshi (1976) point out that vegetable growing gives Indian farmers a better income than other types of farming, and that recent improvements in vegetable varieties are as sensational as those widely publicised with new cereal varieties. A survey by FAO (1971) found that vegetable consumption in India was almost the smallest in the world. It is therefore likely that market gardening will soon increase greatly, and it is fortunate that several by-product leaves have already been studied there. Yields of LP from the brassicas were 90 to 160 kg ha^{-1} (Matai *et al.*, 1973; Deshmukh *et al.*, 1974; Tekale & Joshi, 1976). Other useful by-product leaves from market gardens were chicory (*Cichorium intybus*) (Mahade-viah & Singh, 1968) and sweet potato (*Ipomoea batatas*) (Byers, 1961; Balasundaram *et al.*, 1974a; Deshmukh *et al.*, 1974; Walter *et al.*, 1978), but some varieties are inconveniently mucilaginous (Joshi, 1983) and some contain exceptionally large amounts of phenolic material (Horigome, 1983). Beet (Tekale & Joshi, 1976; Chakrabarti *et al.*, 1984) and radish (Tekale & Joshi, 1976; Chakrabarti *et al.*, 1984) give good yields in India when the leaves are taken at the normal time for harvesting the roots. In Pakistan, Hussain *et al.*

(1968) estimated that 60 000 t of LP could be made from by-product leaves – half of it from radish.

Unpublished experiments at Coimbatore showed that leaves gathered from cotton (*Gossypium hirsutum*) after the second picking of the bolls yielded 510 kg of extracted protein ha^{-1}. After the third picking the leaves were too dry to extract well. Methods now used for mechanised cotton picking depend on chemically defoliating the plants; though it may not be easy to combine collection of leaf with collection of bolls, the area devoted to cotton is so large that the attempt would be worth making. Ramie (*Boehmeria nivea*) is now being grown more extensively than hitherto because of the increased cost of artificial fibres and the exceptional resistance of ramie fibre to weathering and water damage. The leaf is scutched off before retting and contains 2.25 to 4.5% N according to age and level of manuring (Squibb *et al.*, 1954). Judging from appearance and feel it should extract well, and in (unpublished) experiments at Coimbatore 63% of the protein was extractable; less protein was extracted in other experiments (Ghosh, 1967) unless alkali was added. Some other fibre plants have similar N contents but a harsher feel; this is in agreement with the observation (Byers, 1961; Matai *et al.*, 1971) that only 30 to 40% of the protein in sunn hemp (*Crotalaria juncea*) and only 20% of the protein in roselle (*Hibiscus sabdariffa*) is extractable. The latter has an acid leaf and would therefore probably extract better if neutralised. Kenaf (*Hibiscus cannabinus*) was originally grown as a tropical fibre plant; there is now increasing interest in it as a source of paper pulp in New Zealand (Withers, 1973), the USSR and the USA. It has acid leaves and is presumably similar in other ways to roselle. Some of the jutes (e.g. *Corchorus* sp.) have mucilaginous leaves that give an extract that is difficult to handle; this should be remembered lest it be assumed that every protein-rich leaf could be a source of extracted LP.

Protein extraction from other fibre plants is less likely to be successful. Sisal (*Agave sisalana*) is acid. Judging from experience with protein from other leaves, neutralisation after pulp has been made will not make as much protein soluble as would remain soluble if the leaf were scutched with dilute sodium carbonate so that neutralisation and disintegration are simultaneous. There is no reason to think that this neutralisation would damage the fibre. It may also be possible to extract LP from abaca (*Musa textilis*). Byers'

(1961) experience with banana (*M. sapientum*) was not encouraging, but Oke (1983) was more successful, and Romero & Diaz (1984) describe extraction from plantain (*M. paradisiaca*) leaves at pH 11. New varieties of abaca are being tried as sources of paper pulp in several countries. Some may contain less of the phenolics which are the probable reason (p. 23) for poor extraction from members of this genus.

Tobacco (*Nicotiana tabacum* and related species) has been reinvestigated as a source of LP recently (e.g. Kung *et al.*, 1980; Fantozzi & Sensidoni, 1983; Tso & Kung, 1983; De Jong, 1984; Gopalam & Athinarayanan, 1984). It is well known, from long experience with tobacco in virus research, that LP is readily extracted. Tobacco is listed here among the by-products because it is suggested that the fibre residue after protein extraction would still be a desirable material for smoking. However, most of the nicotine would be removed in the 'whey'. Unless that is separated and added back to the residue there seems to be little reason why people, who smoke largely for the sake of the nicotine, should smoke tobacco residue rather than any other form of hay. Sheen (1983) found so little nicotine in the residue that it would be safe as fodder. Working on a laboratory scale his yield of 'cytoplasm' (p. xi) protein was equivalent to 250 to 520 kg ha^{-1} from a single harvest. If that yield can be replicated on a field scale, tobacco has potentialities. The merits of tobacco protein will be discussed later (p. 112).

By the time the seeds are ripe, the foliage of most seed legumes has withered, but there are exceptions. Abu-Shakra *et al.* (1978) found five out of several thousand soya (*Glycine max*) crosses in which the foliage remained green when the seeds were ripe. LP extractability was not measured. If there is rainfall up to the time of harvest, groundnut (*Arachis hypogea*) leaves extract reasonably well (Byers, 1961). Seed yield from some legumes is not diminished when part of the foliage is removed at an early stage (Pal *et al.*, 1982; Pandey, 1983; Pickle & Caviness, 1984). Early defoliation has little effect on the final yield of maize (Hicks & Crookson, 1976).

Increased tillering as a result of removing early growth has been observed with several other species. It was at one time standard practice to allow sheep to graze on cereals in autumn so as to make use of leaf which might be 'scorched' by frost in winter and, it was hoped, to increase tillering in spring. Mechanical harvesting in this manner could yield a useful by-product.

It is often possible to control, or eliminate, weeds by regular harvesting. LP has been extracted from *Parthenium hysterophorus* (Savangikar & Joshi, 1978) and from *Trianthema monogyna* (Mehrotra *et al.*, 1978) which are troublesome in parts of India. Although the merits of a mixture of sunflower and radish have already been mentioned (p. 28), and some other mixtures of conventional crops gave good results (Kasture & Mungikar, 1984), it would be unwise to assume that much use could be made of natural weed mixtures. Furthermore, they are usually unfertilised and grow on rough ground from which harvesting would be difficult. After a phase of nearly universal condemnation, the primitive technique of swidden, or slash-and-burn agriculture is gaining some support (e.g. Dove, 1983). Sometimes, the first growth of weeds, after taking a conventional crop, may be useful.

Tree leaves

When first confronted with the idea of making LP, most people think immediately of tree leaves. Nevertheless, there has been very little work on them. A few points are obvious. The autumn leaf fall is too dry and contains too little protein to be a likely source; extraction is equally unlikely from the hard, dry leaves of many tropical trees; it would usually be difficult to mechanise collection of the leaf from large trees. There are several ways of coping with the last difficulty. As knowledge increases about the normal process of leaf-shedding it should become possible to induce it in the waste branches that accumulate during forestry, and thus separate leaves from twigs. Complete separation may not be necessary. In the USSR, *muka* is made as fodder by beating the branches in a mill and separating the leafier fraction (Young, 1976). Trees could be grown as large hedges and trimmed mechanically from time to time – as is already done with tea, again in the USSR. Coppicing is the most probable technique. It is already used for making paper pulp and silage in the USA (Herrick & Brown, 1967; Dutrow, 1971). Once the cycle is established, the many stems growing from sycamore stumps are harvested mechanically and stripped of their leaves when they are 2 to 5 cm thick. When trees are being selected as sources of pulp, or for 'energy plantations', consideration should be given to the suitability of their leaves for LP extraction: these leaves could become the main source of LP.

Tree and bush leaves are eaten in many countries as green

vegetables, and they are important sources of ruminant fodder. They resemble other leaves in protein content and amino acid composition (Siren *et al.*, 1970; Nedorizescu, 1972; Siren, 1973). Unfortunately, LP seems to be less often readily extracted from them than from the leaves of conventional crop plants (Crook & Holden, 1948). The only abundant species near Rothamsted from which extraction is satisfactory is elder (*Sambucus nigra*). Byers (1961) made LP from *Leucaena*. Extraction is, however, difficult because the extract is gelatinous (Lyon & Kohler, 1981; Mohan & Srivastava, 1981) and tends to coagulate spontaneously. This is presumably the result of association with phenolic material and that may be the reason (Cheeke *et al.*, 1980) for the poor nutritive value of the product. Nevertheless, *Leucaena* (bayani, ipil-ipil, kao haole) deserves fuller study: there are many varieties, it 'fixes' N, grows luxuriantly, and it has limited value as a fodder because of the presence of mimosine, a toxic amino acid, in solution in the juice. Mimosine is not present in the LP (Cheeke *et al.*, 1980). Comments on *Gliricidia maculata* conflict. Balasundaram *et al.*, (1974*b*) and Joshi (1983) report poor extraction: Mohan & Srivastava (1981) were more successful and there was more N in LP from it than in LP from five other tree leaves. Gonzalez *et al.* (1977) made LP from poplar (*Populus* sp.). Several papers contain favourable comments on *Cassia*, but leaves seem sometimes to have been harvested from very young plants. The coagulum from willow (*Salix*) juice contained only 19.5% of true protein and it had poor nutritive value (Näsi, 1983*a*). Jokl & Carlsson (1984) made the interesting observation that LP from *Eucalyptus saligna* poisoned rats unless it had been incubated with *Aspergillus*. Several mammals and marsupials seem to thrive on leaves which are toxic to most species. A study of the intestinal flora of these resistant species would be useful if this phenomenon is common.

Although little is known about LP from tree leaves, some space has been devoted to them because a food-producing tree crop would be the ideal replacement for tropical rain forest. It is now obvious that ecological disaster follows attempts to cultivate annual plants in regions where there is frequent intense rain. Unless soil is protected by the continuous presence of the roots of perennial plants, erosion is a serious risk. Trees usually thought of as replacements for natural rain forest produce an exportable commodity such as rubber or palm oil; it would be advantageous if some trees produced protein for local consumption instead.

Water weeds

Mixed vegetation growing on untended land is not likely to be a useful source of LP (p. 45). The situation is different with water weeds. There is more reason to control their growth because, when excessive, it increases water loss from reservoirs in hot countries, obstructs flow and navigation, and is often a health hazard because the growth harbours disease vectors such as the snails which are part of the cycle causing bilharzia. Much effort is therefore expended on battering water weeds to death *in situ*, killing them with herbicides, or establishing a system of 'biological control'. When excessive growth is caused by sewage or run-off from heavily manured farmland, these methods are only a partial remedy because the elements causing excess growth (eutrophication) are returned to water and promote further growth. When herbicides are used, they or their breakdown products remain in the water and make it unsuitable for irrigation.

Early suggestions (e.g. Pirie, 1960) that water weeds should be regarded as an asset rather than a problem, and that a tiny fraction of the 1000 M£ which is plausibly suggested as the amount spent on ineffective control, should be spent on finding uses for them, were at first treated with scepticism or derision. The suggestion is now being acted on. The US National Academy of Sciences (1976) published a book, *Making water weeds useful;* FAO published (1979) a *Handbook of utilization of aquatic plants* containing about 200 abstracts of relevant papers; a review by Pieterse (1978), unlike an earlier review by him, pays some attention to use; and the 'Aquatic Weed Program' of the University of Florida (Gainesville, USA) organises a computerised survey of world literature and will send references on defined aspects of the subject to those with a genuine interest. All this is very encouraging, but most effort is being put into using water weeds as animal fodder, fibre or fuel rather than as a source of LP.

Processes that drag weeds out of the water, either to be used or allowed to rot, remove nutrients from the water, where they cause eutrophication, on to the land where they increase fertility. The amount of N, phosphorus and potassium removed depends on the luxuriance of growth, i.e. on the amount of N, phosphorus and potassium present, and on the ratio of the concentrations of these elements in the water. Thus Wooten & Dodd (1976) say that water hyacinth removes almost all the N, but little of the phosphorus, whereas Ornes & Sutton (1975) say that it removes 86% of the phosphorus. In one water, N was presumably limiting growth, and in the other water, phosphorus. Boyd (1976) stressed the merits of water weeds as a means of removing from water the elements on which excessive weed growth depends.

Water hyacinth is the most abundant, troublesome and decorative of the weeds. It is said to cover 200000 ha in India, and similar areas elsewhere. The total area is probably more than 1 000000 ha with an annual growth of 10 to 30 t DM ha^{-1}. The DM of the whole floating plant contains about 2.5% N, the separated leaves contain up to 5%. An impressive amount of protein is therefore potentially available. Unfortunately, it does not extract readily unless alkali is added (Ghosh, 1967; Taylor *et al.*, 1971; Matai, 1976). Values for the amino acid composition have been published several times (e.g. Abo Bakr *et al.*, 1984); they fall within the usual LP range.

Extraction is being commercialised (Monsod, 1976) in the Philippines and the LP is used there as pig feed (Alcantara & Lobos, 1981). In Egypt (Borhami & El-Shazly, 1984) it is used experimentally in broiler rations and for buffalo calves. A Commonwealth Science Council Report (1981) makes several brief references to the extraction and use of LP from water hyacinth in Bangladesh, Fiji and Malaysia.

Two other floating weeds, Nile cabbage (*Pistia stratiotes*) and the fern *Salvinia auriculata*, extract as badly as water hyacinth at their natural pH (about 6) (Byers, 1961; Matai *et al.*, 1971; Matai, 1976; Rothamsted Experimental Station, unpublished); the effect of adding alkali to them has not apparently been tried. It would be easy and economical to collect these floating weeds with equipment mounted on a barge, process the extract on it, and discharge the soluble material. The compact extracted fibre and LP would be off-loaded less often than harvested weed would have to be. That manner of working would obviously not control eutrophication as effectively as total weed removal.

Several scientists (e.g. Mills, 1984) and commercial organisations have made, or suggested, mobile equipment for preparing leaf juice. In my opinion, except when a small sample is being made experimentally (p. 18), when a crop growing on land is used, it is easier to move the crop to the equipment, rather than the equipment to the crop. Water weeds are an exception to this principle.

Published figures for the usual growth rate of water hyacinth show that a unit able to process 1 to 2 t hour^{-1} could be kept in daily operation on about 50 ha of infested water.

Protein appears to extract more easily from rooted than from floating water weeds: there is no obvious physiological basis for this distinction. From mixed weeds collected in Hertfordshire (UK), 47% of the protein was extractable (Pirie, 1959c); from a water lily (*Nymphaea lotus*) in Ghana 40% (Byers, 1961), and from one in Alabama (*Nymphaea odorata*) 61% (Boyd, 1968). Boyd (1968, 1971) lists several other species that extract well; *Justicia americana* yielded 300 kg of LP ha^{-1} when harvested in May or June. Some of these water weeds regrow with disconcerting readiness after being cut as a means of control. Unfortunately the reeds (*Typha* and *Phragmites*), which produce very heavy crops in suitable regions (Dykyjova, 1971), do not extract well. Matai (1976) extracted from *Chrozophora* sp. and *Allmania nodiflora* 37% of the leaf N in the form of LP.

No measurements seem to have been made of the extractability of the protein in the algae or angiosperms growing in sea water. If the problem of collection could be solved, marine angiosperms seem to have potentialities. Thus turtle grass (*Thalassia testudinum*) is as productive as most conventional crop plants; its rhizosphere fixes 100 to 500 kg of N ha^{-1} annually (Patriquin & Knowles, 1972); growth is therefore probably limited by phosphorus deficiency. Most of the algae (seaweeds) contain too little protein to make extraction probable. Those species that are rich in protein, e.g. *Ulva* and *Porphyra*, are already used as food.

Miscellaneous unconventional species of leaf

It was at one time reasonable to examine plants found growing wild so as to characterise the types of leaf from which LP could readily be extracted. This enabled a set of expectations to be formulated which could guide the search for species that might be better than conventional crop plants as sources of LP. This was the policy lying behind the work of Crook & Holden (1948) on 26 species, Byers (1961) on 76, Singh (1964) on 12, Devi, *et al.* (1965) on 16, Nazir & Shah (1966) on 25, Gonzalez *et al.* (1968) on 42, Deshmukh & Joshi (1969), Dev & Joshi (1969) and Gore & Joshi (1972) on 61, Sentheshanmuganathan & Durand (1969) on 12, Garcha *et al.* (1970) on 20, Lexander *et al.* (1970) on 24, Matai *et al.* (1971) on

20, Subba Rau *et al.* (1972) on 19, Matai & Bagchi (1974) on 34, Carlsson (1975) on 71, and the species listed between 1948 and 1969 in Rothamsted Annual Reports. Balasundaram *et al.* (1974*b*) examined 212 species, but gave details about the most promising ones only. There is considerable overlap in the lists of plants in these publications: about 300 species have been studied. Various methods were used: yields given in one paper cannot therefore be compared with those in another, and none gives precise information about the yields that would be given by a large-scale extractor. Useful as this work was, it belongs to what may be called the Natural History of the subject. Publication of more measurements of this unsystematic type does not seem to be useful unless the species studied is a member of an Order, or perhaps even a Family, no member of which had already been tried as a source of LP. It is, for example, interesting to know (Dayrell & Vieira, 1977) that LP with the usual properties can be extracted from the cactus *Pereskia aculeata*. That information may give rise to research on more pestilential members of that Order.

Except in such special circumstances, the only measurements which are useful today are those made in a standardised manner, e.g. with the IBP unit (Davys & Pirie, 1969; Davys *et al.*, 1969), or with a comparable unit which has been tested on a crop from which the extractability of LP is well known. When measurements are made on plants picked from odd sites in the neighbourhood, they demonstrate only that a certain species may be worth investigating more thoroughly. Selected species should then be deliberately cultivated on land of known manurial status, harvested at a known age, and the yield from a known area should be measured. Information less comprehensive than that is now scarcely worth publishing.

Of the quarter of a million species of flowering plants, the conventional crop plants may well be the most suitable when they are used in conventional ways and are grown in climates similar to those of the region in which they originated. Many analyses of leaves from wild plants (e.g. Slansky & Feeny, 1977) show that they may contain more N than leaves of cultivated plants grown in the same conditions. They are often more resistant to attack by pests and predators. This resistance is often correlated with the presence of toxic substances, but these would probably be separated from LP during its preparation (p. 60).

Hitherto, we have decided which crops should be grown, and have then extended their range into increasingly improbable regions. The alternative approach is to examine species that already grow well in a region to see whether some use, in this context the extraction of LP, could be made of them. Attention should first be given to species that are already cultivated on a small scale, and that are not used as sources of leaf. For example: quinoa (*Chenopodium quinoa*) is cultivated for its seeds in the Andes, its soft texture makes protein extraction economical (Pirie, 1966*a*), it is a species which has been studied in detail in the laboratory (Carlsson, 1975, 1983; Ostrowski-Meissner *et al.*, 1980), and several papers comment on the excellence of LP made from it. Mustard (p. 25) is cultivated mainly as a green manure in Britain; its relative *Brassica nigra* is an oilseed crop in India. If harvested young it regrows well and gives excellent yields of LP (Matai *et al.*, 1973, 1976). Turnip (*Brassica rapa*) was disappointing; it does not regrow well (Matai *et al.*, 1973) and the yield of LP is poor when it is allowed to grow until there is a saleable root so that the leaves can be considered a by-product (Matai *et al.*, 1976). *Brassica carinata*, which is used as an oilseed crop in Ethiopia, appears to have potentialities in Texas as a source of LP (Brown *et al.*, 1975; Brown & Saldana, 1976). A more unexpected suggestion is that *Tithonia tagetiflora*, usually cultivated because of the beauty of its flowers, should be used; it regrows well after harvesting, and responds vigorously to manuring (Deshmukh *et al.*, 1974; Mungikar *et al.*, 1976*b*). LP from *Galega orientalis* was better poultry feed than LP from bean (*Vicia*) leaves or clover (Näsi & Kiiskinen, 1985).

There seems to be little substance in the argument sometimes advanced that LP would be more acceptable if it were made from leaves already familiar as vegetables. So few communities eat vegetables to the extent which is both desirable and physiologically reasonable (Pirie, 1975*a*, 1976*a*, *c*, 1981*b*, 1985*a*) that nothing would be gained by extracting protein from leaves which could have been eaten in the normal manner by adults. The situation is different with infants. Partial removal of oxalate, present in many leafy vegetables, is advantageous and, when the rest of the diet is bulky, there may not be room in their small stomachs for the quantity of leafy vegetables which would be needed to meet their β carotene (pro-vitamin A) requirement; the equivalent amount of

LP is easily accommodated (p. 123). Familiar green vegetables are no better as sources of LP and β carotene than some species discussed in earlier sections of this chapter. Instead of investigating the extraction of LP from familiar vegetables for infant feeding, it would be better to see whether some of these other species, when cultivated with skilled husbandry, could be used as leafy vegetables exceptionally rich in β carotene, as young lucerne and oats (*Avena sativa*) already are in some places. Nevertheless, Gunetileke (quoted in Pirie, 1971*a*) claims impressive yields from *Basella alba*, and Fafunso & Bassir (1976) extracted *Corchorus olitorus*, *Solanum incanum*, *Solanum nodiflorum* and *Talinum triangulare*. The yields of LP were, however, no greater than those from forage crops.

A special section of this chapter was devoted to water weeds because they grow persistently, to an extent that interferes with the use of the water, and so have to be removed. Weeds grow to a similar extent on land only when that piece of land has for some reason been rejected. Elsewhere they are usually controlled mechanically or with herbicides. There is therefore seldom any reason, in spite of the extraction of LP from *Parthenium hysterophorus* (p. 38), to look upon their use as a means of control. Soil from which weeds are regularly harvested will soon be depleted of nutrients, and regular harvesting will usually alter the composition of a sward. Unless an area is manured and reseeded regularly, harvesting will eradicate some weed species. From this it follows that successful exploitation of a wild plant, or 'weed', will necessitate domestication and cultivation of it as a new crop; cogent reasons will therefore have to be given for cultivating a new species rather than a conventional one.

A few apparent defects of wild plants can be dismissed summarily. Though some grow vigorously, some seem unprepossessing simply because they are growing on poor soil. Even with good husbandry some, which may extract well in the laboratory, may grow slowly. A glance at pictures of the wild ancestors of our crop plants shows what improvements skilled selection can bring about. Uniformity in the time of germination is an important characteristic for which conventional crop plants have been selected. Wild plants tend to be much less uniform: this is an understandable aspect of evolution for it ensures that every individual plant of a species will not suffer should there be inclement weather throughout a region at about the time of germination. Uneven germination should not be considered a serious defect in an uncultivated species because uniformity could probably be introduced quickly by deliberate selection.

Many uncultivated plants are poisonous because they contain alkaloids and similar substances. This also is not a serious defect because alkaloids would be separated from the LP during the process of extraction (pp. 60, 104). Some species, however, are very strongly flavoured and it may be difficult to remove flavours from the LP. In my opinion it is a pity so much of the LP that is now being used is made from lucerne because most varieties have a strong flavour which is not easily removed from the LP. This should be borne in mind by anyone judging the acceptability of LP solely on the basis of experience with lucerne. When made from many other species, LP has little or no flavour. Nevertheless, LP from lucerne is indubitably edible (see Chapter 7); when made from a few other species it is not. An extreme example is *Leonotis taraxifolia* (called the duppy-pumpkin in Jamaica); it grows vigorously, extracts well and gives an unusable product. It is important therefore to assess the flavour of a carefully made preparation of LP at an early stage in the study of any new species.

The primary reason for suggesting that wild plants deserve investigation is that almost all our crop plants were chosen from among wild plants 5000 to 10000 years ago. There is no reason to assume that our ancestors, at the time of what is often called the 'Neolithic Revolution', had unsurpassable skill in identifying potentially useful plants. Furthermore, as techniques of biochemical engineering evolve, it becomes possible to use hitherto useless plants (Pirie, 1962). Another reason, stressed by Hudson (1975, 1976) and by Wareing & Allen (1977), is the meteorological situation in Britain and several other countries. In spring, by the time there is abundant light and moisture, and day-time air temperature is adequate for growth, soil temperature is still too low for the roots of our conventional crops to be able to absorb nutrients. But low soil temperatures, at both the beginning and the end of the year, do not stop the growth of many weeds, e.g. dog's mercury (*Mercurialis perennis*), chickweed (*Stellaria media*), Jack-by-the-hedge (*Sisymbrium alliaria*) and groundsel (*Senecio vulgaris*). For purely scientific reasons it would be interesting to know the basis of this physiological difference between crops and some weeds. It is not impossible that the useful ability to grow in cold soil could then be introduced into some conventional crops. At Rothamsted, nettles (*Urtica dioica*) yielded 612 kg of LP ha^{-1}, this is more than we got in one harvest from any other weed (Pirie, 1959c). Carlsson (1975) put nettles near the top in his survey of 71 species, but concluded that the Chenopodiums were probably the group most worthy of fuller investigation.

It would be unreasonable to devote more than a small fraction of the total effort put into research on LP to the study of unconventional species. Nevertheless, this study would be a useful piece of academic research. Besides the possibility that the period of active photosynthesis could be extended if more plant species were included in the cropping sequence, diversity, either of species growing together or grown in succession, offers many possible advantages. Suitably designed mixtures would contain species that grew when others were dormant, that exploited the full depth of soil for nutrients and water, and that were not all susceptible to the same pests and diseases.

4

Separation, purification, composition and fractionation

The manner in which LP will be used, the manner in which juice was extracted, and the species from which it was extracted, influence the procedures used for separation and purification. If LP is to be used as human food it should be as free from contaminants as possible. The crop should therefore be washed, before being pulped, to remove soil and dust from it. Crops from which LP is made will usually be harvested regularly, if perennial, and after a short period of growth if annual. They will therefore be less subject to attack by pests than conventional crops and there should be less need to use insecticides and herbicides. Use of these potentially toxic substances should be avoided until more is known about the extent to which they are removed by washing leaves. The most convenient and thorough method of washing is to dump the crop into a tank of water, and drag it out continuously under a fine spray of water at a rate suited to the capacity of the equipment used for extraction. Wetting the crop in this way, although it increases the volume of juice handled, increases the percentage of protein extracted. A crop that will be washed must be harvested by a reaper of the old-fashioned type with a reciprocating blade. If it is harvested by a machine with flails or a rotor, and is then blown into a trailer, it will not only be contaminated with soil sucked up by the harvester, but will also be so bruised that much LP will be lost during washing. There is less need to wash the crop when LP is being made as animal feed. Obviously, the point already made (p. 19) about the need to avoid delay between harvesting and processing a crop, applies with even greater force in bulk-extraction. The larger the mass of leaf, the quicker it deteriorates.

Juice made in an arrangement that separates pressing from pulping contains less fibre than juice made in any type of screw expeller because the fibre is not so heavily and assiduously rubbed across the barrel of the expeller (p. 146). Juice from any extraction unit that could be used on a large scale will initially contain an amount of fibre that would be unacceptable in human food. Unacceptability does not arise because fibre is harmful: on the contrary, the ratio of fibre to protein will usually be smaller in LP than in conventional DGLV, and there is a growing body of opinion that many people eat less fibre than is needed for optimal gut behaviour. The main objection to incomplete removal of fibre is that the particles become clearly visible when LP is dried and cause adverse comment. For many reasons (p. 80) LP should be used undried whenever possible. An additional reason is that particles of fibre are then less obtrusive. Fibre is easily removed by passing leaf juice through any of the standard types of rubbed or shaken sieve. A simple rotary arrangement is illustrated in the IBP Handbook (Pirie, 1971a).

Preparation of unfractionated leaf protein

Juices of the species of leaf from which bulk supplies of LP have up to now been made have pHs between 5.6 and 6.4. Little protein would extract from naturally acid leaves unless alkali were added during the extraction, and if alkali is being added, it is essential to add enough to take the pH as far as 8. Otherwise pectin methyl esterase will liberate acid and cause the pH to drift back again (Holden, 1945). Addition of alkali increases the extractability of protein – it also increases the extractability of pectic substances. These coagulate along with the LP. Alkali may be added to increase the stability of carotenoids (p. 80), and alkaline sulfite may be added to inhibit the oxidation of phenolic substances (p. 97). Finally, some leaf extracts are naturally alkaline. Alkaline extracts from a few leaves coagulate satisfactorily on heating, others form a finely dispersed coagulum that is difficult to separate. That problem can be overcome by adjusting the pH. Manipulating the pH of an extract is troublesome, especially when LP production in a village is contemplated. As a rule, for the sake of simplicity, it will be best to forgo the advantages of alkalinity, either natural or induced.

Most of the individual proteins in those leaf extracts that have been studied, precipitate between pH 4 and 5: those proteins that do not precipitate in this range when purified, tend to precipitate along with the others when an unfractionated leaf extract is acidified.

Acidification has been suggested as a method for separating LP in bulk, although curd so produced is so soft and hydrophilic that it is difficult to handle on filter cloth. Furthermore, acidification does not kill many of the potentially pathogenic microorganisms which will inevitably be present in leaf extracts, as in any other material at risk from soil contamination. For the same reason, the suggestion that LP should be allowed to coagulate spontaneously by fermenting the juice anaerobically for a few days, seems even more hazardous. Food Standards officials are often unreasonably suspicious of any novel foodstuff: in this case their doubts would be justified. Fermentation has the added defect that, during it, much of the LP is destroyed by proteolysis. For example, Ameenuddin *et al.* (1984*b*) made, by heat coagulation, LP containing 51% true protein, whereas by fermentation the product contained only 34%. Furthermore, fermentation increases the risk of pheophorbide formation (see next paragraph). All the LP used in extensive human trials was made by heat coagulation. Subba Rau & Singh (1970) got better results in rats with heat-coagulated compared to acid-coagulated LP: Nanda *et al.* (1977) reported contrary results. Other aspects of the fractionation processes used may explain the difference. Saunders *et al.* (1973) found acid-coagulated 'cytoplasm' protein less digestible. If these differences are real, and not caused by, for example, uncontrolled differences in the conditions of drying, it would be interesting to know the reason(s) for them.

All the protein in a leaf extract of suitable pH coagulates in a few seconds at 70 °C. Brief heating is sufficient; sudden heating is desirable because it produces a curd that is hard, granular and easy to manage on a filter (Morrison & Pirie, 1961); it also minimises enzymic changes. All the juice enzymes are not inactivated at 70 °C and they may cause changes in the protein during prolonged storage. With a crop such as lucerne, which contains more chlorophyllase than most crops, heating to 100 °C is advisable. Otherwise, the enzyme splits phytol from some of the chlorophyll and the resulting chlorophyllides readily lose magnesium to yield pheophorbides (Arkcoll & Holden, 1973; Holden, 1974) (p. 56). LP from a lucerne extract that had been heated slowly to 80 °C contained enough pheophorbide to photosensitise rats (Lohrey *et al.*, 1974; Tapper *et al.* 1975). That material also photosensitised pigs (Carr & Pearson, 1974), but a commercial product seems to have been prepared more carefully and caused little trouble (Carr & Pearson,

1976). This photosensitisation is analogous to hypericism caused by various plant pigments and would presumably (Darwin, 1868) affect only animals with white patches and kept in strong light, but it should be avoided if possible. Preparations heated to 100 °C absorb atmospheric oxygen more slowly than those heated to 70 °C (Shah, 1968), but there is so much nonenzymic oxidation when dry material is stored without protection that this difference is probably not significant. The only known disadvantage in heating extracts to 100 °C rather than 80 °C is the extra consumption of steam.

As soon as a leaf is pulped, autolysis begins; the rate at which protein is digested varies with the leaf species (Tracey, 1948; Singh, 1962; Batra *et al.*, 1976) and probably also with the age and nutritional status of the crop. For maximum recovery of protein, there should be no avoidable delay between pulping and heat coagulation; if nucleic acid is considered a harmful component of LP, some delay is advisable so as to allow it to be hydrolysed to fragments which do not coagulate along with the protein (pp. 6, 17). As often, incompatible requirements necessitate compromises.

When working with a few litres of juice, sudden heating is easily managed by heating a pan of *water* to 80 °C and then, with vigorous stirring and continued heating, running in cold juice at such a rate that the temperature in the pan never falls below 70 °C. The pan should have an overflow 5 to 8 cm up its side. The overflow should be 2 or 3 cm wide and about 10 cm long: the pan itself can conveniently be 15 to 30 cm wide. In large-scale work, steam injection is more convenient. A suitable unit for the purpose is a U-shaped, 3- to 5-cm (internal diameter) iron pipe with an outlet for curdled juice and an outlet to cope with surges on one side, an inlet for the steam at the bottom, and the inlet for cold juice at the top of the other leg. This simple arrangement (Pirie, 1957) has been complicated, without being improved, by so many people that it may be justifiable to reprint a drawing of it (Figure 4). Steam is turned on first, the unit will reach 90 °C in a few seconds, juice can then be run in at such a rate that it comes out at the required temperature. If that temperature is greater than 90 °C, the vertical arms of the U-shaped pipe should be extended so as to contain the inevitable surging. A thermostat valve on either the steam or juice inlet is a refinement that is hardly necessary because, if juice comes from a header tank in which the level is kept nearly constant, and if steam pressure is constant, the unit will run at constant temperature with minimal supervision. Every complication that is introduced seems to increase the risk that some underheated juice will get mixed with coagulated juice: filtration will then be troublesome. Larger versions of this unit work satisfactorily: the process of steam injection into a flowing stream is part of conventional chemical engineering.

An argument sometimes advanced in favour of other methods of coagulation is that heating is expensive. Three points should be borne in mind. Because of the risk of microbial contamination, LP will be cooked at some stage: it might as well be cooked at the start. Other methods of coagulation produce a soft curd which is difficult to handle on a filter. The amount of energy needed for coagulation need not be as large as is sometimes thought. Hot 'whey' can be used to preheat juice to 30 or 40 °C without any risk of making a soft curd: in cold weather that would halve the amount of energy needed for coagulation. When the pulper is driven by a steam or diesel engine at the site where LP is being made, no new energy would be needed for coagulation. When an electric motor is used, the 60 to 70% energy loss, inevitably involved with a heat engine, is discarded at the power station. A local prime mover would be less efficient at converting fuel into mechanical energy, but the waste

Figure 4. A convenient arrangement for coagulating leaf juice.

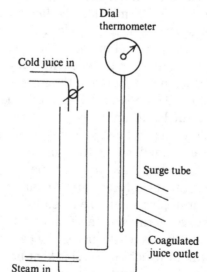

heat is discharged on site and, with the pulpers now being used, it greatly exceeds the amount needed for coagulation.

Air forced out of juice when it is heated makes most of the coagulum float; it can all be made to float by bubbling air (or preferably N_2) through the suspension. Alternatively, the suspension can be stirred gently until the coagulum sinks. These are refinements. If steam injection has been managed competently, the coagulum filters off so easily that little is gained by removing some 'whey' by decantation. Whatever course is adopted, it is advisable to separate the coagulum as quickly as possible because soluble phenolic substances in many leaf extracts combine with the protein if contact is prolonged. This diminishes the N content and digestibility of LP (p. 95).

When juice from less than 300 kg (wet weight) of leaf is being processed, the coagulum can conveniently be collected by pouring coagulated juice into cloth 'stockings' about 18 cm wide (when flat) and 150 cm, or more, long. These are hung up to drip for as long as 'whey' runs out freely. Figure 5 shows the simplest method so far devised for suspending the 'stockings' and closing the lower end. This is also a method which does little damage to 'stockings' in repeated use, and it makes removal of the press-cake easy.

When dripping becomes slow, the 'stockings' are laid between

Figure 5. The simplest arrangement so far devised for hanging a 'stocking' and closing its lower end.

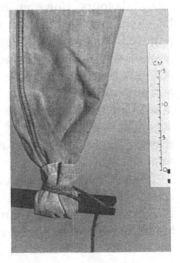

corrugated mats and pressed at 0.3 to 0.5 kgf cm^{-2} (30 to 50 kPa) for a few hours. Rubber mats, with rectangular rather than V-shaped grooves 2 to 4 mm wide, are ideal. An alternative is to wrap plastic strip, of the type used to make chair seats, round a plank. After pressing for a few minutes the LP in a 'stocking' is hard enough for another 'stocking' to be put on top of it – and so on. Obviously, there should be a double-faced corrugated sheet between the 'stockings'. Pressure is most simply applied by means of a weighted lever (p. 146). It is more important to arrange things so that 'whey' can escape freely, than to press heavily, and it is advisable to spread the curd within a 'stocking' so that the final thickness of cake is not much more than 1 cm. To a first approximation (Smiles, 1970), the time taken for a moist slab to reach a required moisture content while under pressure is proportional to thickness^2 × pressure^{-1}. When pressure is applied by means of a weighted lever, although doubling the area occupied by a given amount of curd halves the pressure on it, it still halves the time needed for pressing because the slab is thinner. It is easy to press to 50% DM and possible to press to > 60%. With larger amounts of leaf, some form of belt press, pump-fed filter press, or decanting centrifuge will be needed. Coagulated leaf juice has no peculiar properties which necessitate modifying standard chemical engineering practice with such equipment.

Dry leaf powders are used as flavouring agents and relishes in many parts of the world; there is therefore no universal dislike for 'leafy' flavours. However, during the initial phase of introducing LP as a human food, as much flavour as possible should be removed. Furthermore, some potential sources of LP, e.g. potato haulm, contain water-soluble materials that are slightly toxic; others, e.g. lucerne, contain growth depressants (Peterson, 1950); and all contain reducing sugars. The details of the Maillard reaction, in which lysine is made unavailable by combination with reducing sugar, have been studied by Finot & Mauron (1972). For all these reasons as much soluble material as is practicable should be removed from LP.

If 'whey' contains the usual amount of soluble material, this will contribute 7.5% to the DM of a sample of LP pressed to only 40% DM. There will be still more in less adequately pressed material. Similarly, if a well-pressed coagulum is resuspended in so much water that that contains 1 g or less of soluble material l^{-1}, the final press-cake should contain only about 0.1% soluble material. The recommended standard is < 1% of soluble material (Pirie, 1971a).

As already mentioned, the coagulum from a suddenly heated extract filters off easily; when resuspended in water at leaf pH, filtration is difficult. Filtration is easy if salt is added to the water or if the suspension is acidified to about pH 4. It is not essential to measure this pH. Different samples of press-cake are so similar that it is safe to assume that 200 ml of 2 N acid will be sufficient for 1 kg of cake containing 40% DM, and the pH is usually not critical. When, as in experimental work on the properties of LP, it is important to know the pH, particular attention should be paid to crumbling the cake finely and it should be soaked for several hours at the required pH. Even then, a sample of curd, apparently at pH 4, if ground finely in a mortar and resuspended, often has a pH as great as 4.5. For rigid control of pH it is advisable to adjust the pH of the initial press-cake when that has been pressed so briefly that it contains only 30% DM and therefore resuspends easily. It will then be necessary to use a larger volume of wash-water, or to resuspend and press twice.

Besides facilitating filtration, resuspension in an adequate amount of acidified water removes alkaloids to a large extent, and produces a product with the keeping qualities of cheese, pickles or sauerkraut. Unfortunately, acidification converts the chlorophylls completely to pheophytins through loss of magnesium (p. 56). The dull green is less attractive to most people than the bright colour of neutral LP, and the risk of pheophorbide formation is increased. There is more loss of carotene from acid than neutral LP and the unsaturated fatty acids in it are more rapidly oxidised. If LP is regarded as something to be made on the farm, these defects of acid washing must probably be accepted. If it is regarded as an industrial product, it could be washed by centrifugation or, if washed by acidification and filtration, the washed, pressed and crumbled coagulum could be neutralised again by exposure to ammonia vapour.

Composition of unfractionated leaf protein
Components soluble in lipid solvents

The DM of LP, made in the manner described above, contains 20 to 30% of material soluble in the usual lipid solvents. The amount of lipid is often underestimated because, although the material is largely soluble in ether or petroleum ether after it has been extracted, only part of it is released from combination with other components of LP by these solvents. The remainder can be extracted by a mixture of chloroform and methanol, or by a mixture of alcohol and ether in the presence of a little strong acid (Holden, 1952;

Figure 6. (*a*) Chlorophyll breakdown chart. (*b*) The structure of chlorophyll A. Definite positions are not assigned to the double bonds in the main ring. The fluidity of the bonding in that structure is probably part of the reason for its effectiveness in photosynthesis.

(*a*)

(*b*)

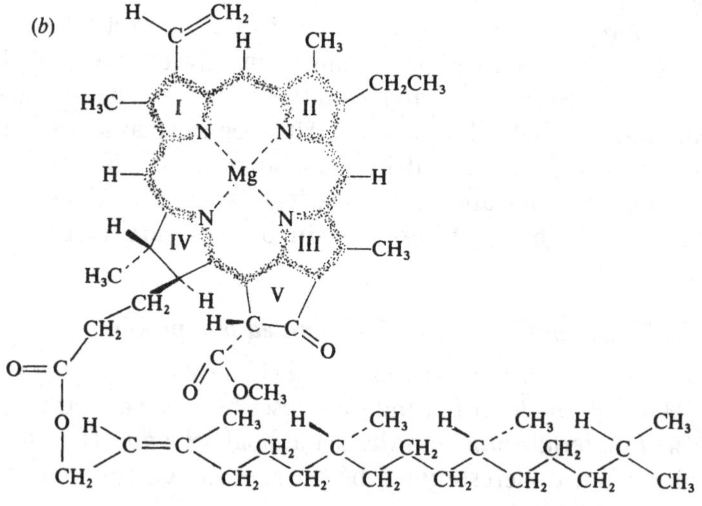

Buchanan, 1969*a*, *b*; Byers, 1971*b*). The most noticeable, but nutritionally less important, component is the mixture of chlorophylls and their breakdown products (Figure 6). The importance of processing leaf extracts in such a manner that pheophorbide is not formed has already been stressed (p. 50). That hazard is easily avoided. Properly made LP, still containing the green pigments, should probably not be used as food by people with Refsum disease, a rare enzyme anomaly in the metabolism of phytanic acid derived from phytol (Herndon *et al.*, 1969; Masters-Thomas *et al.*, 1980), nor by the small group of people with an intestinal flora which makes the photosensitising pigment phylloerythrin from chlorophyll (Clare, 1952). Obviously, such people are well advised to avoid DGLV also.

Nutritionally and commercially the carotenoids are the most important lipid components of LP. Xanthophyll is commercially valuable because it makes chicken's legs and skin yellow (Colker *et al.*, 1948); this colour is admired in several countries. Trivial as this may seem, it is the main reason for commercial production of LP at present (1986). β carotene (pro-vitamin A) is nutritionally important because vitamin A deficiency, leading to xerophthalmia and blindness, is widespread (p. 124). Freshly made LP contains 0.5 to 2.0 mg β carotene g^{-1}. Most of it is in the 'chloroplast' fraction; the content therefore depends on the extent to which the technique used to make leaf juice brings out chloroplasts and their fragments, on the initial chloroplast content of the leaves, and on the carotene content of their chloroplasts. There is more β carotene in upper, sunlit leaves than in inner or lower leaves, and more in leaves taken at dawn and dusk than at midday. The extensive literature on these points is surveyed elsewhere (Pirie, 1984*a*, *d*). Later sections (pp. 80, 124) discuss the likelihood that β carotene is often underestimated in dried LP, its stability during storage, and the part that β carotene in LP could play in human nutrition.

Fatty acids account for most of the 20 to 30% of lipid in LP. Although LP is suggested primarily as a protein supplement, its energy content is therefore not negligible – it would, for example, be greater than that in most cuts of meat. No trustworthy figures for the energy available *in vivo* from LP have been published. Several recent papers e.g. Lima *et al.* (1965), Buchanan (1969*a*, *b*), Hudson & Karis (1973, 1974), and Nagy & Nordby (1983), confirm and extend the old observation that leaf lipids are rich in doubly and

Palmitic	$CH_3 (CH_2)_{14} COOH$
Palmitoleic	$CH_3 (CH_2)_5 CH : CH (CH_2)_7 COOH$
Stearic	$CH_3 (CH_2)_{16} COOH$
Oleic	$CH_3 (CH_2)_7 CH : CH (CH_2)_7 COOH$
Linoleic	$CH_3 (CH_2)_4 CH : CH CH_2 CH : CH (CH_2)_7 COOH$
Linolenic	$CH_3 CH_2 CH : CH CH_2 CH : CH CH_2 CH : CH (CH_2)_7 COOH$

Figure 7. The dominant fatty acids in leaf protein.

trebly unsaturated fatty acids. The dominant acids are palmitic, palmitoleic, stearic, oleic, linoleic and linolenic (Figure 7). Although the unsaturated acids usually account for half the total fatty acid there are marked apparent differences in their ratios in different LPs. It is at present not clear what importance should be attached to these differences because the fatty acid pattern of leaves changes with age (Eichenberger & Grob, 1965; Kannangara & Stumpf, 1972; Hudson & Karis, 1974) and with the temperature at which plants are grown. There is therefore no reason to postulate genuine species differences. The point deserves thorough study because of the nutritional importance of unsaturated fatty acids (Elliott & Knight, 1972; Whitehead, 1980), and because their presence in LP complicates preservation and storage. Hudson & Karis (1974) point out that some leaf crops produce more lipid ha^{-1} in a shorter time than most oilseed crops.

Carbohydrates

The amount of starch in LP depends on the species used and the weather at the time of harvest. There is so much starch in extracts from some leaves, notably peas taken on a sunny day, that grains collect as a white sediment if the extract is allowed to stand. Unless the starch is centrifuged out of the extract before coagulation, 5% (or even more) will be present in the LP. Even when no starch grains are visible in the extract, there is usually 5 to 10% of carbohydrate in LP. By graded hydrolysis followed by paper chromatography, Festenstein (1976) showed that this carbohydrate differs in composition from the carbohydrate mixture in 'whey' filtered from the coagulum. That is to say, the presence of this carbohydrate is not entirely a consequence of the LP being inadequately washed. When alkali is added during the extraction, the carbohydrate content is increased because of the presence of pectic substances. Nucleic acid supplies a little of the carbohydrate in most preparations (p. 17).

Miscellaneous components

If LP has been thoroughly washed and pressed it will contain less than 1% of water-soluble material. Consequently it will be nearly free from B vitamins and other such components of the leaf: their presence (Bray, 1976) is evidence of inadequate washing. LP will, however, contain useful amounts of vitamins E and K (p. 81) besides β carotene.

The ash content varies. It is seldom less than 3%, and may be as great as 8%. Dust from the surface of the leaves is the source of much of it, but some of the grasses contain silica which is partly extracted along with protein. Silica is usually considered harmless: there is even the suggestion (Schwarz, 1977) that it is beneficial. Recently however (Parry *et al.*, 1984; O'Neill *et al.*, 1986), evidence has accumulated that certain forms of silica can be cancer promoters. Among the forms of silica incriminated are those in the seed hairs of *Phalaris* and *Setaria* and the smoke from burning sugar cane. In any probable diet, LP is likely to supply only a small fraction of the total silica intake, but the form of silica in leaves of different species may be another factor that should be taken into account in choosing sources of LP.

Cations, such as Ca^{2+} and K^+, are to a large extent removed during washing at pH 4; that is near the isoelectric point of most of the proteins that make up LP. An IBP Technical Group (Pirie, 1971*a*) suggested that when LP is to be used as human food it should contain < 3% ash and < 1% acid-insoluble ash. Later experience suggests that it may not always be possible to meet that standard. Nevertheless, material containing more ash than that should be carefully examined to see whether the ash is an unavoidable component of LP made from the species that is being used, or whether it is present because of some deficiency in the technique of preparation.

Plants grown on some soils absorb deleterious metals, such as lead and zinc, and translocate them to their leaves. When leafy vegetables are eaten in the conventional way, little attention is paid to the presence of lead in them though amounts as large as

50 mg kg^{-1} of DM have been recorded. Although lead is known to be more abundant in the chloroplasts than in other parts of the leaf (Holl & Hampp, 1975) much of the lead in leaves may be the result of surface contamination with earth, and the plants may have come from roadside sites where they are fouled by lead from motor car exhaust. There is apparently no legal restriction on the amount of lead that may be present in a 'natural' food; restrictions apply as soon as a food is processed. This may restrict the sites where crops for LP production can be grown unless some method, such as coagulation in the presence of a chelating agent, can be found that limits the amount of lead accompanying the LP.

The presence of alkaloids in a leaf species is unlikely to be an obstacle to its use as a source of LP because all alkaloids are soluble on the acid side of neutrality; they will probably be removed when LP is washed at pH 4. This expectation must obviously be verified for each species. Other components, such as saponins in lucerne and estrogens (p. 103) in some clovers (Glencross et al., 1972), may present more problems. Fortunately, there are varieties of most cultivated species that contain only small concentrations of these substances. Until the risk has been shown to be negligible, any crop suffering from a heavy fungus attack should be rejected lest a mycotoxin should be carried through into LP. This last risk is not peculiar to LP; many edible seeds and tubers, especially after storage, may be contaminated with mycotoxins.

After extraction with lipid solvents, and after allowing for the presence of carbohydrate and ash, bulk preparations of LP invariably contain less N than is usually found in proteins. Phenolic compounds and other tanning agents are present in all species of leaf; sometimes they are a major component and seriously interfere with protein extraction (p. 23). There is evidence (Davies et al., 1975, 1978; Newby et al., 1980) that phenolic acids are linked to some of the leaf protein *in vivo*. They accompany the protein during fractionation even when stringent precautions, such as grinding the leaf under liquid N, are taken to avoid secondary combination. However, most of the phenolic material found in LP is probably acquired during extraction and processing. Jennings et al. (1968) measured the absorption spectrum of bean (*Vicia faba*) LP, compared it with the spectrum calculated from its tyrosine and tryptophan content, and showed that the difference between the spectra resembled that of phenolic material. Extreme values for phenolic material in LP from nine species were 1.4 to 2.2% (Subba Rau *et al.*, 1972) and 0.8 to 4.0% (Maliwal, 1983). Phenolics were therefore not major components of any of these LPs. Analytical results are not always easy to interpret. For example, Donnelly *et al.* (1983) found that extraction in the presence of sulfite increased the amount of phenolic material extractable with ethanol – presumably because sulfite had suppressed covalent binding.

Separation of phenolic material from LP after isolation is an obvious alternative to preventing complex formation. This is possible by extracting with 10 M urea (Barbeau & Kinsella, 1983), with mercaptoethanol in the presence of 2.5% sodium chloride (Lahiry *et al.*, 1977), or with acetyl bromide (Rambourg & Montes, 1983). Such procedures shed light on the nature of the union between LP and phenolics, but they are not suggested as practical means for disrupting that union.

In the course of a thorough review of reactions which occur, or may occur, between phenolic compounds and LP, Pierpoint (1983) commented '...the reaction between phenols and proteins are highly appreciated since they will add significantly to the flavor, taste and appearance of products such as tea, coffee, beer and tobacco. It is conceivable that in the future the phenolic content of some preparations of LP will be similarly appreciated for its contribution to their taste.' In the meantime however, apart from the possibility (p. 82) that they protect β carotene from destruction when LP is stored, phenolics must be considered harmful. Relative freedom from them will be one of the criteria used in choosing plants for LP production even though they seem to protect plants from predation.

Phenolic compounds tend to increase with maturity or stress (e.g. Milic, 1972; Wong, 1973). When studying the extent to which they impede protein extraction, and damage what protein is extracted, some methods used in the laboratory to promote the extraction of enzymes or viruses should be tried. For example: air could be excluded during extraction (Cohen *et al.*, 1956; Pirie, 1961), a reducing environment could be maintained with —SH compounds (Hageman & Waygood, 1959; Anderson & Rowan, 1967), or substances such as nicotine could be added to compete with protein for conjugation with the phenolics (Thung & Want, 1951; Thresh, 1956). More recently, polyvinylpyrrolidone has been used in many operations in which it is desirable to sequestrate tanning agents.

When added to grass or lucerne extracts before heat coagulation the amount of phenolic material in the LP was halved (Fafunso & Byers, 1977). Methods such as these could probably not be used in large-scale work. Sulfite is cheap enough to be used in making bulk preparations. Many experiments (pp. 97, 100) show that, when added during pulping, sulfite improves the appearance and nutritive quality of LP.

Amino acids

To varying extents, amino acids are destroyed when proteins mixed with carbohydrate and lipid are hydrolysed with acid. Byers (1971*b*, 1983) described the precautions that should be taken to minimise destruction. In spite of these precautions, even experienced analysts (e.g. Williams *et al.*, 1979; Walker, 1982) report considerable scatter in the results of replicated analyses. Results are particularly unsatisfactory with cyst(e)ine and methionine. This is unfortunate because these are the amino acids in which LP is most likely to be deficient. Because of this analytical uncertainty, it would be well if editors insisted on evidence that amino acid analyses had been done two or three times, preferably on different protein samples, before allowing the analyses space in a journal. Space should certainly not be wasted on analyses of samples of LP which have been dried in an oven without preliminary washing to remove 'whey' components. Any differences found are more likely to be the result of Maillard reactions than to be genuine species differences. Furthermore, relatively trustworthy analyses show such uniformity (e.g. Bryant & Fowden, 1959; Pleshkov & Fowden, 1959; Gerloff *et al.*, 1965; Byers, 1971*a*, *b*; Fafunso & Byers, 1977) that there is little need to publish further similar results.

Uniformity in amino acid composition is, at first sight, surprising because there are great differences in the specific enzyme activities of leaves from different species, or differing in age. However, apart from RuBP, each enzyme accounts for such a small fraction of the total LP that, even although they have different amino acid compositions, the average composition is little affected by changes in enzyme ratios. These considerations lead to the conclusion that one of the less necessary pieces of equipment to install in an institute where work on LP is starting is an amino acid analyser. If a species should be found that extracts well and is promising from an agronomic standpoint, perhaps from an Order from which LP has not hitherto been made, it is likely that some institute already

Table 2. Amino acid composition (expressed as weight of amino acid (g) 100 g^{-1} recovered amino acids) of three types of leaf protein preparation from three species. Ammonia, cyst(e)ine and tryptophan are excluded (From Byers, 1971a)

Amino acid	'Chloroplast'			Unfractionated			'Cytoplasm'		
	Barley	Lupin*	Chinese cabbage†	Barley	Lupin	Chinese cabbage	Barley	Lupin	Chinese cabbage
Aspartic acid	9.75	10.10	9.86	9.57	10.22	10.01	9.62	10.01	10.05
Threonine	4.82	4.97	4.87	5.07	5.01	5.22	5.41	5.04	5.43
Serine	4.85	5.15	5.24	4.40	4.68	4.50	4.10	4.13	4.12
Glutamic acid	11.00	11.35	11.28	11.41	11.88	11.91	11.94	12.15	12.21
Proline	4.88	5.06	4.98	4.68	4.79	4.72	4.62	4.79	4.46
Glycine	6.12	5.97	6.53	5.64	5.69	5.35	5.38	5.32	5.29
Alanine	7.05	6.40	6.45	6.71	6.21	6.10	6.52	5.99	5.91
Valine	6.16	6.10	5.64	6.37	6.27	6.06	6.50	6.32	6.17
Methionine	2.28	1.90	2.11	2.24	1.70	1.94	2.39	1.76	2.13
Isoleucine	5.25	5.76	5.03	4.95	4.93	4.62	4.74	4.42	4.38
Leucine	10.43	10.68	10.40	9.33	9.75	9.29	8.42	9.21	8.79
Tyrosine	4.49	4.20	4.19	4.50	4.61	4.71	4.92	5.56	5.07
Phenylalanine	6.97	7.16	6.85	6.22	6.24	6.24	5.84	5.82	5.87
Lysine	5.60	4.78	5.23	6.61	6.60	7.08	7.06	7.30	7.23
Histidine	1.82	1.91	2.00	2.34	2.31	2.42	2.66	2.82	2.65
Arginine	6.29	6.13	6.33	6.89	6.35	6.46	7.01	6.67	6.89

* Lupinus albus
† Brassica chinensis

equipped with an analyser will be willing to analyse one or two samples. In the unlikely event that the LP has an unusual composition, more detailed cooperation can be organised.

Table 2 (from Byers, 1971a) illustrates the reason for expecting uniformity in amino acid composition. The differences between the three species, and between the unfractionated LP and the 'chloroplast' and 'cytoplasm' fractions (p. xi) made from it, hardly lie outside the range of unavoidable uncertainty in amino acid analysis. This uncertainty arises not only from errors in measuring the areas under the peaks on the elution profile given by an automatic analyser, but also from the fact that if less than 100% of the N in the sample is accounted for in the amino acids recovered there is no guarantee that losses are distributed among the amino acids in the same way in all samples. Nevertheless, the small differences in amino acid composition are probably real. This was demonstrated by an elaborate statistical study (Byers, 1971b). Essentially, the difference between the largest and smallest content of each amino acid was found for the whole group of 39 samples. Then, the amino acid contents of each sample were compared in turn with the amino acid contents of each of the other 39 minus 1 samples. The criterion of comparison was the extent to which the difference in each amino acid content in each pair was less than the difference for that amino acid in the whole group of 39. Because cyst(e)ine and tryptophan are partly or wholly destroyed during acid hydrolysis, this was done for only 16 amino acids. It nevertheless produced a formidable collection of figures. The next process may metaphorically be described as arranging the figures in 39 minus 1 dimensional space and instructing a computer to search for the plane in that space that contained the largest differences in amino acid composition. Figure 8 shows the 39 samples plotted in accordance with their two coordinates in that plane. Although the three types of preparation, from three species, are not separated completely it is clear that there is some grouping into nine categories.

Gerloff *et al.* (1965) included tryptophan in their set of analyses. If attention is restricted to preparations containing more than 60% protein, all the values are in the range 1.3 to 2.2%; when results are expressed as in Table 2, that comes to 2 to 3 g of amino acid 100 g^{-1} of total amino acids. The scatter is not unexpected; tryptophan determinations are notoriously unsatisfactory, and some heavily contaminated samples of LP gave suspiciously large values. Cyst(e)ine is also not included in the table because of uncertainty about the extent to which it is destroyed during hydrolysis in the presence of phenolic substances (Fafunso & Byers, 1977 and p. 61) and of substances that form humin. Destruction is complete if air is not excluded. Even with careful exclusion of air, Byers' (1971*b*) values varied from 0.69 to 3.04 g of cyst(e)ine per 100 g of recovered amino acids in the nine types of preparation – a range very much larger than that found with any other amino acid. Well-washed LP that has been thoroughly extracted with lipid solvents is not known to contain any sulfur in forms other than cyst(e)ine and methionine. An upper limit for the cyst(e)ine content can therefore be deduced from the difference between the total sulfur and the sulfur contained in methionine (Subba Rau *et al.*, 1972; Byers, 1975). Byers analysed 32 preparations; by direct analysis the percentage of cyst(e)ine in the true protein component of LP from lucerne was 0.82 to 2.15, and by calculation 2.35 to 2.78. With lupin LP the corresponding values were 'trace' to 2.79 and 2.00 to 2.67. Obviously, the validity of the method depends on

Figure 8. Statistical analysis of the amino acid composition of 39 preparations of leaf protein. By a technique outlined in the text, the analyses were arranged by a computer in a manner designed to demonstrate the reality (if any) of differences in composition between preparations of different types. The numbers within the figure have no quantitative significance, they merely identify preparations.

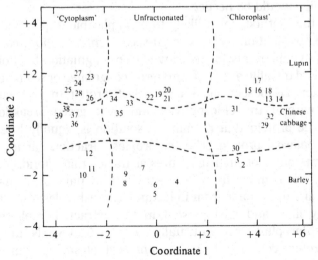

the validity of the assumption that sulfur is never present in LP in a third form.

Fractionation of leaf protein

Alongside purely academic work on the fractionation of the proteins in leaf extracts, there has been a great deal of work designed simply to remove those parts of the protein mixture which are associated with chlorophyll; it is sometimes assumed that a green product will be less acceptable as human food than one coloured brown or fawn. The validity of that assumption will be discussed later (p. 100). A similar problem confronted those making enzyme or virus preparations in the period before high-speed centrifuges holding 100 ml or more had become common pieces of equipment. Chloroplasts and their fragments were then coagulated by adding calcium salts, phosphate, ethanol or other solvents miscible with water, or weak acids (e.g. Bawden & Pirie, 1938). Methods such as these are advocated from time to time for bulk production. Partly because of expense, but mainly because of the varied composition of leaf juice, they are not likely to be practical. For example, calcium salts are presumably effective when juice contains excess phosphate, and phosphate when juice contains excess calcium. The concentration of such components depends not only on species, but also on age, fertiliser treatment and weather. A prearranged routine using these methods can be effective with plants cultivated in controlled conditions: with field-grown material, a preliminary study of each batch of juice would be needed.

Among the methods most likely to be applicable in practice are differential coagulation by heat, high-speed centrifugation, coagulation by solvents immiscible with water or coagulation by ionised polymers, and ultrafiltration. The material coagulated or sedimented by these agents is loosely called 'chloroplast' protein (p. xi) because it carries with it the chlorophylls and their green breakdown products; the protein that remains in solution is, equally loosely, called 'cytoplasm' protein. The 'chloroplast' fraction carries out with it components of such structures as nuclei, mitochondria and ribosomes; when made from leaf extracts that have been inadequately strained, or made from inadequately washed leaves, it also carries any fibre and dust present in the extract. For all these reasons, 'chloroplast' protein usually contains 6 to 8% N, unfractionated protein contains 9 to 11%, and 'cytoplasm' protein may contain 16%. 'Chloroplast' protein is less digestible (p. 86) both *in vitro* and *in vivo*; the quality of the remaining protein is therefore improved if it is removed. These are real merits, and when industrial production starts it is possible that it will be worthwhile separating pale 'cytoplasm' protein because it could be presented in food in more ways than unfractionated LP. But no method of separation is likely to be simple enough to be a domestic or farm operation. Furthermore, the β carotene is in the 'chloroplast' fraction. It will be a pity if it is discarded as animal feed in countries where there is a shortage of vitamin A (p. 124) in human diets.

Most of the protein in a leaf is present initially in the chloroplasts. Even the refined methods used to separate them in the laboratory release some of their contents. The rougher methods used in bulk extraction probably cause still more damage. One of the factors that distinguishes species classified as good sources of LP is probably chloroplast fragility, because the smaller the fragments into which chloroplasts break the more likely they are to be disentangled from leaf fibre. One advantage of pulping with added water may be that osmotic shock when chloroplasts are released into dilute suspension disrupts them and so releases more of their protein. Electric pulses have been used to disrupt similar cells, e.g. erythrocytes (Zimmermann *et al.*, 1975). An electric pulse technique which may depend on this process is advocated by 'Licensintorg' in the USSR for improving LP extraction. The idea deserves thorough study.

When pale 'cytoplasm' protein is required, chloroplast fragility is a defect because chlorophyll-containing fragments seem often to be less easily sedimented or coagulated than intact chloroplasts. For example, in spite of the exceptional stability of quinoa chloroplasts in some circumstances (p. 5), Extracts from frozen quinoa leaves were not readily cleared of chlorophyll by high-speed centrifuging (Lundborg, 1980*a*). By contrast, leaf extracts were often frozen in the 1940s as a means of increasing the sedimentability of chloroplasts and their fragments when virus preparations were being made. Because of this uncertainty in its effect, freezing is not likely to become a useful technique in bulk fractionation of LP.

Fractionation by heating

Rouelle's (1773) method of heating in two stages (p. 155) is still the most widely used fractionation procedure. Coagulation is so rapid that even momentary local overheating must be avoided. A few hundred millilitres of juice can be swirled in a flask heated over a flame. On a larger scale it is better to maintain one side of a heat-exchanger at the required temperature, and to scrape continuously the side over which the juice flows (Pirie, 1964b; Kayama, 1984), or to arrange for violent turbulent flow against it (Christensen, 1984). The curd has then to be collected by centrifuging: it is too soft for recovery on a filter.

Byers (1967b) found 61 to 64% of the protein in lupin juice in the 'chloroplast' fraction. The figures were 80% for rape juice and 92% for nasturtium (*Tropaeolum majus*) juice. These measurements were made at the natural pH of the leaf extracts, and at 43 °C with lupin but 53 °C with the other two. Working at 53 °C throughout, Lexander *et al.* (1970) adjusted the pH of extracts from 14 species to values between 4.5 and 6, and found, as others have done, that the proportion of the protein separating in the 'chloroplast' fraction increased as the pH diminished. With extracts from frozen sunflower leaves, all the protein was in that fraction even at pH 6; the smallest amount of protein was in that fraction with nettles, lucerne and *Atriplex hortensis*. This work was extended by Carlsson (1975) who heated extracts from 41 species to 53 °C at their natural pHs. The more alkaline leaves produced the least 'chloroplast' coagulum: of those leaves that extracted well, the ones giving the smallest amounts of 'chloroplast' protein were beet (32% at pH 7.1), spinach (43% at pH 6.5) and *Atriplex latifolium* (50% at pH 6.6). Using a more extensive range of temperatures, times of heating, and pHs, Lundborg's (1980c, d) results were similar but differed on a few points. It is clear that there can be large differences between extracts from different species, or between extracts from the same species when the conditions of heating and centrifuging differ.

'Cytoplasm' protein coagulates when juice from which 'chloroplast' protein has been removed is heated to 80 or 100 °C. After pilot-plant studies (de Fremery *et al.*, 1973; Edwards *et al.*, 1975), the process was patented (Bickoff & Kohler, 1974) in spite of the long period during which it has been in common use. To avoid loss of coagulable protein, caused by the action of proteolytic enzymes in the leaf extract, the extract must be heated quickly to about 60 °C, held at that temperature for only a few seconds and then cooled quickly before being centrifuged. In the arrangement used at Rothamsted (Pirie, 1964b), the outflow from the scraped heat-exchanger passed through a metal tube cooled with ice. The supernatant fluid then has to be reheated. It may be possible to manage these manipulations economically, but it would not be prudent to assume this. The first phase of the separation has, however, considerable academic interest. For example, Jones & Mangan (1976), using lucerne extracts to which inhibitors of oxidative processes had been added, removed the 'chloroplast' fraction and then separated a homogeneous 'cytoplasm' fraction with the physical properties of RuBP.

The preparation of partly dried fibre, that can be economically conserved as winter fodder by complete drying, and that is not greatly depleted of protein, is likely to be one of the objectives of fodder fractionation (p. 126). This can be achieved by heating the leaf to destroy its osmotic control and then pressing without rubbing or disintegration. The technique was tried 40 years ago (cf. Pirie, 1977) and has been investigated again (Mathismoen, 1974; Gastineau, 1976). In these trials almost all the protein was coagulated *in situ*. Attempts to coagulate the 'chloroplast' protein, while leaving the 'cytoplasm' protein extractable, by heating intact leaves to lower temperatures and then pulping, were unsuccessful (Harendranath & Singh, 1980). When processes such as these are used with leaves containing chlorophyllase (e.g. lucerne, p. 50) there is a risk of pheophorbide formation in the fibre residue unless it is heated or dried soon after the separation.

Centrifugal separation

The feasibility of separating 'chloroplast' protein centrifugally on a commercial scale depends entirely on choosing crops with chloroplasts that do not break up into very small fragments, and on the availability of robust centrifuges that can be run economically at the required speed. There is nothing novel in the principle. After 1940, major improvements in centrifuge technology were immediately used in plant virus research. There is reasonable agreement (e.g. Bawden & Pirie, 1938; Lugg, 1939; Pirie, 1950, 1955; Pierpoint, 1959; Chayen *et al.*, 1961; Wilson & Tilley, 1965; Byers, 1971*b*; de Fremery *et al.*, 1973) that chloroplasts and their fragments sediment at 8000 to 50000 *g* depending on the duration of centrifuging and the depth of fluid through which a particle has to travel to be effectively sedimented. In a centrifuge with swinging buckets, this depth equals the length of the tube; in an angle-head centrifuge, it is not much greater than the width of the tube; in a continuous-flow centrifuge, it is a millimetre or less. To sediment all the green material it is usually necessary to centrifuge at 100000 *g*. However, Lundborg (1980*b*) separated nearly all the green colour from juices from several varieties of *Brassica oleracea* at pH 7.5 and 15000 *g*, about half the protein sedimented as well. With *Atriplex hortensis* 150000 *g* was needed (Lundborg, 1980*c*).

The material removed by high-speed centrifuging is similar to, but not identical with, the material removed by heating to 50 or 60 °C; the supernatant fluid after centrifuging usually still contains some protein that coagulates when heated to 60 °C. If it should prove to be feasible on a commercial scale, centrifuging will be a valuable pretreatment in making good-quality LP from some species. Thus LP made from potato leaf juice that had been centrifuged at 25000 *g* contained 11 to 12% N, whereas without the treatment it contained only 9 to 10% (Carruthers & Pirie, 1975).

Fractionation with solvents

Material similar to the 'chloroplast' protein that coagulates on gentle heating is precipitated more easily than the remainder of the protein by solvents such as methanol, ethanol and isopropanol. But 10 to 20% of the solvent has to be added and recovery of this would be uneconomic. Tsuchihashi (1923) coagulated the cell walls in a suspension of hemolysed erythrocytes by shaking with chloroform. This process was improved by adding water-immiscible solvents, such as butyl alcohol, to the chloroform and, before the advent of gel-filtration, it was used extensively to remove almost all the protein from solutions of other macromolecules (e.g. Heidelberger *et al.*, 1936). By suitable choice of solvents and duration of shaking, the method can be used to coagulate some of the component proteins in a mixture while leaving others in solution. As might be expected in view of the extent to which they resemble erythrocytes, chloroplasts and their fragments are readily coagulated by many water-immiscible solvents. Slade *et al.* (1945) and G. Singh (1984) used amyl alcohol; Allison (quoted in Hove & Bailey, 1975) and Reddy & Joshi (1984) used butanol. Some of the halogenated hydrocarbons have been used in similar processes; they may be preferable in the laboratory, but they are more expensive. Furthermore, if the 'whey' that runs away from the heat-coagulated 'cytoplasm' protein is to be used as a culture medium for micro-organisms (p. 134), there are advantages in using a solvent that can be metabolised by microorganisms and that acts as a coagulant at a concentration small enough not to inhibit their growth.

Ionised polymers

Knuckles *et al.* (1980*a*) compared the efficacy of 54 commercially available substances, normally used to clarify waste water from the food-processing industry, as flocculants for 'chloroplast' protein. Ten were effective. A cationic polymer with molecular mass 100 000 to 300 000 worked when 2 g was added to a litre of lucerne juice. An anionic polyacrylamide (Superfloc A 150) worked at 1 g l^{-1} (Humphries & Bray, 1979). This flocculant was used by Fiorentini *et al.* (1984) at 20 mg l^{-1} and is presumably preferable to 'Prodefloc' which the same group (Anelli *et al.*, 1977) had used earlier. These agents are considered harmless by the US Food and Drug Administration; it may however be legitimate to wonder what the long-term effect of feeding animals on 'chloroplast' protein containing them would be. If used judiciously there will presumably be little contamination of 'cytoplasm' protein by them.

Alum

Hanczakowski & Skraba (1984) used alum to precipitate most of the protein from lucerne juice: Savangikar *et al.* (1985) used it to precipitate the 'chloroplast' protein, and then made 'cytoplasm' protein from the supernatant by heating it, adding more alum, or taking it to pH 4. There is a body of medical opinion opposed to introducing alum into foodstuffs, and exposure to alum increases the lability of β carotene when moist LP is stored (p. 82).

Ultrafiltration

Chibnall's (1939) method of removing the larger particles, including 'chloroplast' protein, by filtering leaf extracts through a 3-cm-thick pad of paper pulp, has occasionally been used by others (e.g. Lugg, 1939; Davies *et al.*, 1952; Yemm & Folkes, 1953; Chibnall *et al.*, 1963). In the original form, it is satisfactory for making 'cytoplasm' protein, but the 'chloroplast' fraction would be difficult to recover in a usable form. Technical developments in the production of smooth-surfaced membranes of graded porosity, that can be scraped clear of 'chloroplast' protein when they get clogged, may make such a method practicable on a commercial scale.

'Cytoplasm' protein in solutions that have been freed from

'chloroplast' protein by any of the methods described above can be concentrated five- to ten-fold by ultrafiltration (Tragardh, 1974; Knuckles *et al.*, 1975, 1980*b*; Ostrowski-Meissner, 1983*b*). Brown material, salts and sugars are largely removed from the resulting concentrate. Although polysaccharides probably remain in the concentrate, curd made by heating or adding acid or ethanol contained more than 93% protein. It is therefore claimed that this procedure gives a purer product than can be made by conventional heat fractionation.

A generalisation of fractionation methods

Ultrafiltration and centrifugal fractionation depend on the difference in average particle sizes and densities of the components of the 'chloroplast' and 'cytoplasm' fractions. All other methods depend on the greater instability of 'chloroplast' suspensions. In most leaf juices, that fraction coagulates when juice is frozen and thawed, left for a few days at about 0 °C or a few hours at room temperature. At 40 °C, less than an hour is usually needed (Singh & Singh, 1980). When LP is precipitated at pH 4, the 'chloroplast' fraction is irreversibly altered so that it is mainly the 'cytoplasm' fraction which redissolves when the precipitate is resuspended at neutrality. Merodio *et al.* (1983) suggested that this procedure could be useful in practice.

The food industry finds bland, colourless, soluble proteins more attractive than an insoluble green protein. In spite of problems raised by the variability of leaf juices, work is therefore likely to continue on methods for making 'cytoplasm' protein. An effective method will probably be so complex that the product will be as expensive as a comparable protein made from a legume seed. That will rob fodder fractionation of one of the merits claimed for it. Most of the protein in leaf juice is in the 'chloroplast' fraction. If that is used only as animal feed, fodder fractionation would be robbed of another of its merits. The technical complexity of the methods needed to separate leaf juice into fractions prevents them from being usable in domestic or village conditions. It is in these conditions, rather than the conditions catered for by the food industry, that novel sources of food protein are most needed. It is unfractionated LP that can be useful in them.

5

Preservation, storage and modification

Unfractioned preparations of LP have often had different nutritive values in spite of having comparable N contents. These differences are probably more often the consequence of differences in the techniques used in preparation and preservation, than of differences in the ages or species of the leaves. As with other foods, all the reactions which take place in LP during preservation are likely to be harmful. Whenever possible therefore, it should be used fresh, in the form of the moist press-cake. But preservation will often be necessary if LP is to be used at a distant place, or in a season when crops do not grow.

Preservation of leaf juice

As liquid feeds gain popularity in commercial pig farming, the idea of using unfractioned juice as, or in, the feed has obvious attractions. It eliminates the need for filtration equipment, the cost of drying, and the possibility that LP will be damaged if drying is mismanaged. Equally obvious disadvantages are the spontaneous separation of curd from leaf juice, and its varied composition. Depending on the maturity of the crop, on the weather, and on the time of day of harvest, the protein content of the juice can vary by a factor of five and its DM by a factor of three. Furthermore, some unevenness in the supply of juice is unavoidable. Apart from spontaneous coagulation, these difficulties can be circumvented by storing juice for a few weeks. The objective would not be to retain the original character of the juice, but merely to retain a character acceptable to pigs. By adjusting the pH to 3 and adding 1 to 2 g of sodium metabisulfite l^{-1} (Cheeseman, 1977 and many other papers suggesting similar treatments) the more obvious forms of spoilage are postponed for many weeks. Näsi (1983b) tried the agents commonly

73

used in silage preservation (acetic, formic, phosphoric, hydrochloric and sulfuric acids, and formaldehyde) and found that all were effective in preserving leaf juice when used at 0.5%. They preserved moist LP at 1%. If there is some proteolysis during storage, the feeding value of the juice is not necessarily diminished because the amino acids are still there and are presumably as valuable as when in the original protein. However, sterility is not complete. There is therefore a risk that nitrite, mycotoxins and bacterial toxins will be made; pheophorbide formation (p. 50) is also a possible hazard.

In spite of these defects, preserved leaf juice has been used in many trials with calves, pigs and poultry. There have been failures – for predictable reasons. The concentration of cations in leaf juice overloads pig's kidneys, and they do not like the flavour of juice from some species, or after a few weeks' storage (Carton & Maguire, 1983; Maguire *et al.*, 1983). Juice has, however, been used success-fully as a supplement in pig diets (Maguire & Brookes, 1972, 1973; Naumenko *et al.*, 1975, 1977; Houseman & Connell, 1976; Ohshima & Ueda, 1982). Prasad *et al.* (1977) and Grover *et al.* (1980) used it with calves. It is not as good as skim milk, but it is cheaper (Rangeekar *et al.*, 1979).

An alternative approach (Pirie, 1977) would be to exploit the coagulation, which accompanies preservation with acids, instead of regarding it as a nuisance. The volume of settled coagulum will depend on the initial protein content of the juice, but 5 to 10% of the wet weight of the settled coagulum is protein. This makes it a more convenient component of a liquid feed than juice. The supernatant fluid, if syphoned off before there has been much proteolysis, is probably more useful as a fertiliser, or as a substrate for microbial growth, than as feed for nonruminants. This partial separation would remove much of the flavour and would diminish the cation load. The advantages of using moist LP, from which much of the 'whey' has been removed, in pig and poultry feeding are now gaining recognition (Maguire *et al.*, 1983; Ameenuddin *et al.*, 1984*a*). As with any perishable fodder, it is obviously necessary to move material through the system quickly.

Preservation of moist leaf protein

When LP is to be used as human food, the original heat coagulum should be suspended in water and pressed again to remove, to the greatest extent possible, flavouring materials as well as carbohydrates which would damage the LP by Maillard (or 'browning') reactions (p. 54). Filtration is facilitated by bringing the suspension to pH 4. This brings it into the pH range of cheese, pickles and sauerkraut, and gives it the keeping qualities of these foods. For more prolonged storage it can be canned. One attempt at preservation by exposure to 2 Mrads of ionising radiation from a cobalt source resulted in material with an unacceptable flavour. It is claimed that this 'wet dog smell' can be prevented if air is thoroughly removed from the food and if the food is frozen when irradiated.

Simpler methods for preserving the moist cake become more satisfactory the more completely water has been pressed out. When pressing, it is more important to arrange conditions in such a manner that the expressed juice can flow away easily, than to apply heavy and prolonged pressure (p. 54). Fully pressed material contains < 50% water. If such material is mixed with one-seventh of its weight of salt, rammed into a jar, and protected from access of air in the ways familiar in jam-making, it keeps well. Each gram of LP DM is then accompanied by about 0.28 g of salt. This is not an unreasonable amount of salt in a food, provided the other components of the day's diet are not also heavily salted. The salt can, however, be partly removed before cooking by soaking the mass in water and letting the LP settle. If the final food is to be sweet, sugar can be used in the same way (p. 81); press-cake keeps well if it contains approximately equal weights of LP DM and sugar. Subba Rao *et al.* (1967) preserved acidified cake for several months in the presence of 2% residual acetic acid and 0.2% orange peel oil. They examined the microbial population of the material in some detail. Coagulation by heating kills 99% of the bacteria initially present in leaf juice; 22 samples of LP contained no *Salmonella* spp. and so few *Escherichia coli* and *Clostridium perfringens* as to be nonhazardous (Sarathchandra & Boyd, 1980). In LP from seven species preserved for 6 months with 1% acetic acid (Arkcoll, 1973*a*) the soil fungus *Mucor racemosus* was the usual contaminant. This initial contamination is unavoidable. To avoid further contamination, the techniques of dairying are needed. Production of LP has much in common with milk production: there is a dirty side to the work and a clean side. The two must be kept separate. Other aspects of preservation with salt and acetic acid are discussed later (p. 81) in connection with the β carotene in LP. Benzoate and sorbate have not, as yet, been favoured as preservatives.

Foods are often preserved by encouraging a desirable growth instead of preventing all microbial growth. Khandelwal *et al.* (1984) found that the flavour of lucerne LP was improved by 6 months' storage after inoculation with three species of *Lactobacillus*. Slade *et al.* (1939) suggested inoculating LP so as to make 'cheese'. This is a possibility because many highly esteemed cheeses have, for the uninitiated, peculiar flavours; but a liking for exotic cheeses is not quickly established. In cooperation with the National Institute for Research in Dairying, acceptable samples of ordinary cheese containing 5 to 10% of LP were made. If LP were being preserved in this way, it would be worth considering the addition of *Streptomyces* to the inoculum so as to produce vitamin B 12 – this is often lacking in vegetarian diets.

Methods of drying leaf protein

Freeze-drying is the least harmful and the most expensive method of drying LP; it would be used only for the preparation of analytical samples, or to measure the amount of harm being done by drying in other ways. For example, Ohshima (1985) compared oven-dried and freeze-dried LP supplemented with methionine as sole protein sources for rats. Only the former needed supplementation with lysine. Fully pressed LP gives a rather hard and granular product when freeze-dried; a more attractive product is made if the cake is moistened so as to make a paste containing 70 to 75% water. The more quickly it is then frozen, the better the texture of the product; it is lumpy when large ice crystals are allowed to grow. By evaporative cooling in a suitable unit (Pirie, 1964c) a layer of paste 1 cm thick can be frozen solid in 1 to 2 min. Freeze-dried material appears to keep indefinitely if stored in an airtight container but, because of its enormous surface, the unsaturated fats in it absorb oxygen rapidly if it is exposed to air. Exposure to light increases the rate of oxygen uptake. Lea & Parr (1961) concluded that, in spite of the heat treatment to which LP had been exposed, enzyme reactions were responsible for part of the oxygen uptake. It is inhibited by the usual antioxidants. LP made from leaf juice that had stood for 2 hours at room temperature before coagulation oxidised faster than LP coagulated quickly; Arkcoll (1973a) suggested that this was the result of destruction of tocopherol (vitamin E), an antioxidant normally present in leaves. Hudson & Warwick (1977) and Hudson & Mahgoub (1980) suggested that more powerful antioxidants must also be present in LP.

When LP containing < 50% DM is dried completely in a current of air, either at room temperature or in an oven, the product is dark and gritty. A more attractive product can be made by drying in a tumble-drier until the water content is only about 30% (this point is easily judged by feel) and then grinding finely before drying is completed (Arkcoll, 1969). LP which has been pressed throughly in a thin layer is sufficiently dry to be treated in this way. The performance of some commercial drying units was examined by de Fremery *et al.* (1972): Straub *et al.* (1979) found that drying on a drum heated to 120 °C caused little damage.

The appearance, and probably the nutritive value, of LP can be improved by drying in the presence of various extenders. Salt is one of these; but if salt is being added it might as well be added as a preservative to moist cake. LP will finally be mixed with some form of flour or meal before use as human or animal food. These substances usually contain 10 to 14% water when in equilibrium with a humid atmosphere, but they can be parched in an oven with little loss of nutritive value. By mixing LP press-cake containing about 40% water with parched cereal products, Duckworth & Woodham (1961) and Foot (1974) made feed mixtures, which did not deteriorate for several months. Using cereal meals with the normal water content and adding propionic acid, Braude *et al.* (1977) made pellets which kept well in spite of the presence of 20% water.

The extent to which LP is damaged when dried in an oven depends not only on the temperature of the air to which it is exposed, but also on the extent to which the LP had been washed to remove sugars, and on the rate at which water is being removed. The more rapid the evaporation, the greater the temperature difference between LP and air, and the shorter the time for which LP is moist when it is being heated. Buchanan (1969b) found that digestibility *in vitro* diminished more when LP was heated when it contained 7% water than when it contained 3%: from this it follows that a small amount of crumbled LP, dried quickly in a thin layer in a current of air, is likely to be damaged less than a larger amount dried in the form of lumps in a closed oven, even though the latter is being dried at a lower temperature. This is probably the explanation of the apparent conflict between the observation (Duckworth & Woodham, 1961) that LP was damaged by drying at temperatures above 82 °C, whereas Subba Rau & Singh (1970) found no damage during drying at 100 °C in a fast current of air. The precise conditions of drying were not stated by Munshi *et al.* (1974) who found, using the growth of rats and the amount of xanthine oxidase in their livers as criteria, that cowpea LP was damaged more by drying at 80 °C than at 60 °C. The consequences of spray drying, or roller drying, incompletely washed LP or unfractionated leaf juice are hard to interpret. Although Cowlishaw *et al.* (1956a, b) had already demonstrated the poor nutritional value of dried whole juice for chicks, this technique was advocated by Hartman *et al.* (1967). The inadvisability of using the technique was stressed by Subba Rau *et al.* (1969) and Pirie (1971a): many components of 'whey' that would be included in the LP have doubtful nutritional value for nonruminants, and some of them are toxic.

Drying by solvent extraction

Solvents that are miscible with water, e.g. acetone and the lower alcohols, can be used for removing water and, at the same time, removing some lipids, and chlorophyll and its breakdown products. This policy is advocated by those who make the curious assumption that a food should never be green. Although removal of the green colour is marginally advantageous, removal of the lipids is detrimental. They contribute nearly as much to the energy content of an average sample of LP as the protein contributes, and the carotene is a valuable component that would be lost (p. 123). A further disadvantage of solvent extraction is that removal of the solvent at the end of the process is nearly as difficult as the direct removal of water would have been. When acetone is used, it is almost impossible to remove the smell of mesitylene and its derivatives completely. Nevertheless, extraction with ethanol is advocated as a prelude to drying (Mori *et al.*, 1984; Ohshima & Moriyama, 1985) because of improved appearance, digestibility *in vitro* and nutritive value. The LP used in these experiments had apparently not been washed; it is possible that resuspension in water would have been equally beneficial, and lipids would not have been lost.

Solvent extraction has, on the other hand, some compensating advantages. The removal of unsaturated lipids, even if it is incomplete, makes it less necessary to exclude air from material that will be stored, and thorough solvent extraction can restore digestibility to LP that has been damaged by prolonged heating when moist (Shah *et al.*, 1967; Buchanan, 1969*a*, *b*). Differences in the extent to which earlier treatments had modified the protein may explain the contrasting results of Murai *et al.* (1980, 1982); solvent extraction improved nutritive value in the first but not the second experiment. Carlsson (1980) suggested that solvent extraction may improve nutritive value by removing saponins.

As Rouelle observed (p. 156), colour becomes less readily extractable after LP has been left for a few days at room temperature. This is true of the lipids also; presumably there is a reaction comparable to the tanning of 'chamois' leather. Buchanan (1969*b*) measured the progress of fixation at 60 °C in the presence of 1 to 2.5 % water. Because of this fixation, and because of the risk of microbial attack, solvent extraction, if it is decided on, should start within a few days of pressing. The lipids in dry soya LP do not seem to become less extractable after storing at room temperature (Betschart & Kinsella, 1974*a*). Mokady & Zimmermann (1966) compared products made by extraction with methanol followed by chloroform, acetone followed by petrol ether, and boiling toluene, used first to distil off the water azeotropically, and then as a lipid solvent; the first system gave the most attractive product. There is general agreement (e.g. Huang *et al.*, 1971; Bray *et al.*, 1978) that polar solvents extract lipids most effectively from LP.

Carotenoids in preserved LP

Solvent extraction partially removes carotenoids from LP and is therefore unsuitable when importance is attached to the xanthophyll in it as a colouring agent for poultry (p. 57), or to the β carotene as a source of vitamin A. Vitamin A (retinol) is not known to occur in leaves. β carotene is abundant and is split and oxidised in animals to retinol and retinene (Figure 9). Retinene is a component of visual purple: its absence causes night blindness.

In theory, one molecule of carotene yields two of retinol. In practice, scission is not quantitative and absorption is incomplete. Taking these factors into consideration, WHO/U.S.AID (1976) recommended 0.38 mg of β carotene kg^{-1} body weight for children and about one-fifth of that amount for adults. Requirement is therefore in the 2 to 6 mg range.

Carotenoids are exposed to enzymic oxidation as soon as a crop is pulped. That action is mainly associated with leaf fibre (Arkcoll & Holden, 1973); juice should therefore be separated as quickly as possible. The pH optimum is on the acid side of neutrality (Walsh & Hauge, 1953; Arkcoll & Holden, 1973); carotenoids are therefore more fully preserved if the pulp is made slightly alkaline by exposure to ammonia vapour early in processing (Spencer *et al.*, 1971). Delay between extracting and coagulating juice is unlikely in commercial production. Half the β carotene may be lost in 24 hours at room temperature (Arkcoll & Holden, 1973; Tekale & Joshi, 1977). There was little further loss during coagulation and pressing. Because of these factors, and because of contamination with iron from the equipment used for making LP, differences in the β carotene contents of the final products are as likely to be caused by differences in preparative technique as by genuine and consistent species differences.

For long-term preservation, and in commercial production, LP will probably have to be dried. This is unfortunate because drying is expensive, usually produces a gritty product, and may destroy part of the β carotene. In some published figures (including some of those quoted here) the extent of destruction may be exaggerated. β carotene is readily extracted by solvents, such as a mixture of acetone and petroleum ether, from moist or freeze-dried LP. Even after fine grinding, complete extraction from oven-dried LP may take 2 or 3 days. Extraction is quicker from LP dried at 30 to 40 °C or after soaking oven-dried material in water for a few days.

Figure 9. The structure of β carotene, retinal or retinene and retinol or vitamin A.

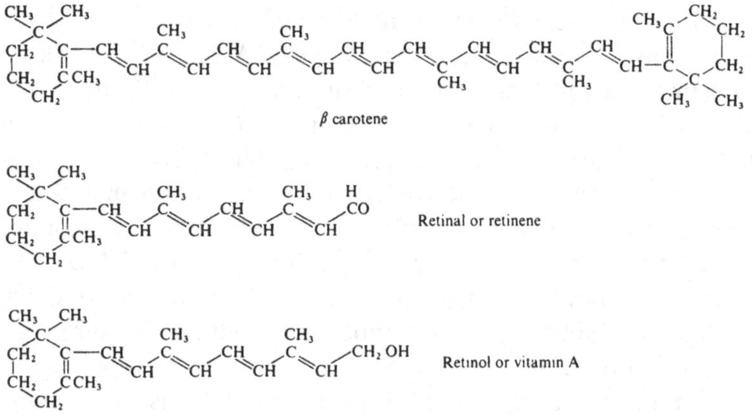

β carotene

Retinal or retinene

Retinol or vitamin A

Soaking has a similar effect on the rate of digestion by proteases (Buchanan, 1969a) (p. 88). During digestion *in vivo* that slowly-extracted β carotene presumably becomes available.

Several commercial drying methods were compared by Miller *et al.* (1972). Retention of β carotene after drying depends on the extent to which LP is kept cold and protected from light and air. After a year at 20 °C in air, 90% was lost; with exclusion of light and air, 12%; there was still less loss at −20 °C (Arkcoll, 1973a). Witt *et al.* (1971) and Gibson & Wallace (1980) found different rates of destruction – presumably because of the variables already listed, and because of differences in the extent to which 'whey' components had been removed. Ben Aziz *et al.* (1968, 1971) and Livingstone *et al.* (1980) comment on the protective effect of water-soluble antioxidants. There are also lipid-soluble antioxidants. Foxell (1977) found 0.36 mg tocopherol (vitamin E) g^{-1} in freeze-dried LP and, in agreement with Arkcoll (1973a) who had found 0.1 mg g^{-1}, less after other methods of drying. Hudson & Warwick (1977) and Hudson & Mahgoub (1980) argue that there would have to be three times as much tocopherol present if it were the only antioxidant. The other lipid-soluble vitamins do not seem to have been studied, nor has the effect of working throughout with stainless steel or plastic equipment.

LP made in semi-commercial conditions contained 0.57 mg β carotene g^{-1} (Miller *et al.*, 1972). When made in the laboratory it contained 0.8 to 1.7 mg g^{-1} (Arkcoll, 1973a). Carefully made material often contains more than 2 mg g^{-1}. There should therefore usually be enough β carotene in 2 or 3 g of freshly made LP to meet the daily vitamin A requirement of an infant, and enough in 4 to 6 g to meet that of an adult.

Brief preservation is often all that is needed – for example, to avoid having to make LP every day, or to cover unavoidable interruptions in the supply of crops. Smaller amounts of preservative than those already listed (p. 75) can then be used, e.g. acetic acid at 7 g l^{-1} or salt at 200 g l^{-1} in the water phase of the press-cake. When sugar is the preservative, each gram of water needs 3 or 4 g (Pirie, 1980). In the dark and at room temperature, but in the presence of air, there is little loss of β carotene in a few weeks with acetic acid. With salt as the preservative, loss is quicker. LP from rape is an extreme example: half the β carotene is lost in 2 days (Pirie, 1984a). β carotene is more stable, in the presence of salt, in

LP from other species, and the actual rate of loss tends to be greater in LP from young than old leaves (Pirie, 1984d). Half is likely to be lost in a week. Loss can be prevented by excluding air, by the presence of inhibitors such as cyanide, or by exposing LP to oxalate, ascorbate, some mercaptans and phenolic compounds. Perhaps because of the presence of phenolic compounds in them, the 'whey' from species such as elder, and extracts made by boiling the leaves remaining after tea has been made in the normal manner, protect β carotene in salted LP. Precipitation with alum diminishes the amount of β carotene in LP (p. 71): exposure to alum reverses, to a great extent, all these protective effects (Pirie, 1986).

This thermostable, salt-activated catalyst is of some academic interest; it is also of practical interest because it seems to be active in salt-pickled vegetables. Further study of the process may lead to a method of inhibition which could be used in practice. The results quoted here were obtained with LP made by heat coagulation. LP made by acid coagulation contains less β carotene initially and it is less stable in moist preserved press-cake.

Xanthophyll is more abundant than β carotene in LP initially, and it is destroyed more slowly during storage. Commercially, it is a very important component (p. 57, 142).

Modification

After being made insoluble at neutrality by heat coagulation, LP is still partly soluble at pH 9 or above. Exposure to the level of alkalinity that is needed to get much of the protein into solution is well known to damage other proteins by racemising some amino acids, by converting alanine into dehydroalanine which then conjugates with lysine and cysteine, and in other ways. Treatment with alkali has, therefore, not been seriously studied. Partial acid hydrolysis is an alternative, but this destroys tryptophan and often produces peptides with strong flavours. This is a defect with enzymic digestion also.

Enzymic digests of some proteins, after being concentrated and, if necessary, fractionated, can undergo some resynthesis or rearrangement to yield 'plastein'. The nature of the reaction is obscure, but it can be used to incorporate into the 'plastein' those amino acids that are scarce in the original protein (Yamashita *et al.*, 1976). The potentialities of this procedure were examined by Savangikar & Joshi (1979). It should be possible by this means to make material containing a greater percentage of protein than the original LP because the digest can be filtered to remove insoluble materials.

Because the final product is insoluble, small molecules that may have unwanted flavours should be removable. An inversion of this digestion procedure is suggested by Staron (1975). Inadequately washed LP was incubated with a fungus (*Geotrichum candidum*) which has little protease activity but contains other active enzymes which would, it was hoped, remove carbohydrates, phenolic compounds and saponins. Staron presents some evidence for the removal of much of the ash and of a protease inhibitor from lucerne LP, and for an improvement in nutritive value. The carbohydrate content was not affected and the amount of LP recovered is not stated. There is as yet no evidence that adequate washing would not have been as beneficial as fermentation. Eucalyptus LP became more palatable to rats, and less toxic, after digestion with *Aspergillus* sp. (Carlsson *et al.*, 1984; Jokl & Carlsson, 1984).

One factor leading to the insolubility of many proteins, including coagulated LP, is the approximate equality between the dominant polar groups, —COOH and —NH$_2$. The properties of a protein can therefore be altered by chemically masking one or other type of polar group. Franzen & Kinsella (1976) made a slightly soluble product by treating LP with enough succinic anhydride to cover 84% of the —NH$_2$ groups. The yields, digestibility and nutritive values of these products have not been reported.

Modified forms of LP will have an enhanced appeal in the conventional food industry because products could be made that disperse easily in water and give textures that are considered desirable. Any such procedure would, however, greatly increase the cost of the product and make production, of necessity, an industrial process. The primary merit of LP, that it can be made economically on a small scale and locally, will be lost if it is extensively modified.

6

Digestibility in vitro *and nutritive value in animals*

Leaves contain both proteases and their antagonists. The former predominate as is shown by the conversion of protein N to nonprotein N in silage and stored leaf juice. That conversion also shows that LP is susceptible to proteolysis. It is reasonable to expect LP, like other proteins, to become more susceptible to proteolysis after denaturation unless later stages of the preparative procedure, e.g. drying and exposure to phenolic compounds, diminish its susceptibility. Leaf proteases have been extensively studied (e.g. Tracey, 1948; Singh, 1962; Ragster & Chrispeels, 1981); they cause little trouble in LP production because they are inactivated during heat coagulation and may be removed when the curd is washed. The protease inhibitors, on the other hand, like the more potent inhibitors in legume seeds, are often relatively thermostable (Chen & Mitchell, 1973; Richardson, 1977). They can be almost completely removed from LP by thorough washing (Humphries, 1980), but they are present in LP made in the standard manner; this was shown by the inhibition of tryptic digestion of casein when LP was added. Humphries found more inhibitor in LP from lucerne than in LP from fescue (*Festuca pratensis*), ryegrass or quinoa, and none in LP from kale; it is present in LP from several tropical species (Carlsson *et al.*, 1984). It is not clear to what extent inhibitors are harmful: there is little correlation between the amount in LP and its nutritive value or digestibility *in vivo*.

Digestibility in vitro

Enzymes of the digestive tract on leaf protein

If a protein is readily digested by several enzymes *in vitro*, it is reasonable to expect that it will be digested in the gut; if it is not digested *in vitro*, it may still be digested *in vivo* because of the simultaneous action of several proteases and of possible cooperation from the gut flora. Nevertheless, measurements of the rate and extent of enzymic digestion are a useful prelude to measurements *in vivo*.

Akeson & Stahmann (1965) digested 18 samples of LP from nine species with pepsin followed by trypsin, removed undigested material with picric acid, and measured the individual free amino acids; 10 to 33% of the amino acids were hydrolysed from 12 conventional food proteins, whereas 19 to 24% were hydrolysed from the LPs. A similar experiment with larger amounts of enzyme (Bickoff *et al.*, 1975) can be interpreted as showing that 88% of the N in LP and 97% of the N in 'cytoplasm' protein appeared as free amino acids, but it is not clearly stated that hydrolysis went so nearly to completion. These may be values for soluble N rather than amino acid N. From the point of view of nutrition this matters little; peptides as well as amino acids are absorbed.

There is no uncertainty about what was measured in the experiments of Lexander *et al.* (1970) and Carlsson (1975) – it was the percentage of the N brought into solution by digestion, and also the percentage of that solubilised N still precipitable by TCA. Using minute amounts of pepsin alone, Lexander *et al.* (1970) found large differences between 23 species. *Amaranthus caudatus* and *Vicia faba* were the most readily solubilised (65%) and half of the soluble N could be precipitated by TCA. They found a clear trend: the larger the percentage of N solubilised, the larger the percentage of it that remained precipitable by TCA. This relationship also held for products of fractionation. 'Chloroplast' fractions were less digestible, and little of what was brought into solution could be precipitated by TCA, whereas digests from the more digestible 'cytoplasm' fractions were 60% precipitable by TCA. It is possible that this phenomenon could be turned into a useful practical method for making more refined fractions from LP.

In comparisons between 15 species, Lexander *et al.* (1970) found that solubilisation by pepsin followed by trypsin ranged from 84 to 40% and little material remained precipitable by TCA. In a similar comparison (Maliwal, 1983), the digestibility of LP from nine species ranged from 22 to 87%; it correlated reasonably well with relative absence of bound polyphenol and carbohydrate. LP from crops given abundant N fertiliser tends to contain more N than LP from less-well-manured crops; there is general agreement (e.g. the papers already cited, and Horigome & Kandatsu, 1964) that the digestibility of LP increases with its N content. Lexander *et al.* (1970), and Carlsson (1975), found a small effect with three species.

There is no need to quote the many papers showing that 'cytoplasm' protein is more readily digested than 'chloroplast' protein, and that it is rarely as readily digested by pepsin and pancreatin as casein. There is unanimity on these points. Digestibility can be improved by changes in the preparative technique. 'Cytoplasm' protein, when made by membrane filtration, was more digestible than when made by heat coagulation; it was still more digestible when made in the presence of reducing agents such as ascorbic acid or sulfite (Ostrowski-Meissner, 1980*b*, 1983*b*). Protein which had been extracted at pH 11 was more digestible if separated at pH 4 without heating, rather than at 5 with heat (Nanda *et al.*, 1977).

Papain on leaf protein

Measurements of digestibility by enzymes from the mammalian digestive tract have an obvious bearing on the potential value of LP as a food for nonruminants. Measurements of digestibility by papain have academic interest and also a bearing on the possible use of modified LP by the food industry. Papain is used extensively and the pawpaw or papaya, from which it is made, is a common fruit in countries where LP could be useful.

Unactivated papain digests LP very slowly; even when activated by potassium cyanide, it digests LP at only one-fifth of the rate at which it digests casein (Byers, 1967*a*). This slow digestion is probably not caused by the presence of antiproteases in LP from the species used because these samples of LP, unlike those studied by Humphries (1980), did not interfere with the digestion of casein. Byers used thoroughly washed LP because, when cyanide is used to activate papain, residual reducing sugars form cyanohydrins which yield ammonia during Kjeldahl digestion (Buchanan & Byers, 1969). Cyanohydrins are soluble; they do not therefore cause trouble if the course of digestion is followed by measuring the amount of undigested protein only. It may, however, be prudent to use thioglycolic acid as the activator (Buchanan & Byers, 1969; Saunders *et al.*, 1973; Betschart & Kinsella, 1974*b*) although it is not so effective as potassium cyanide. As in digestion with pepsin and trypsin, part of the LP is made soluble while remaining precipitable by TCA; the amount of material of this type increases as the pH of papain digests is increased (Byers, 1967a). Papain digestibility of freeze-dried LP is not increased by extraction with lipid solvents at neutrality and it is diminished by extraction with solvents in the presence of strong acids. This effect is probably caused by hardening and compaction of the particles of LP so that there is less opportunity for access by the enzyme. Buchanan (1969a) restored digestibility by fine grinding and prolonged pre-soaking. Papain, like pepsin and trypsin, digests 'chloroplast' protein more slowly than 'cytoplasm' protein whether the fractions are separated by differential heat coagulation (Byers, 1967b) or centrifuging (Byers, 1971b). Species and age differences in the proportions of the various components of LP may explain some of the small differences in digestibility that Byers found when she compared preparations from 14 species harvested at different ages.

Enzymes on heat-damaged leaf protein

Samples of LP that have been subjected to more prolonged heating than is necessary for coagulation, have invariably been digested more slowly than samples from the same batch not so treated. The extent of the change depends on conditions of heating. Obviously, duration and temperature are important variables. The extent to which air penetrated the mass of LP seems also to be important, as is the period during which the LP was heated in the presence of water. As has already been pointed out (p. 77), a thin layer of moist, crumbled LP, heated in a current of air so that water vapour can escape quickly, probably suffers less than lumps of moist LP in a closed oven even although the oven temperature is lower. Finally, there is the point to which attention cannot be too often directed – inadequately washed LP, heated in the presence of extraneous components of the leaf 'whey', has more opportunity for undergoing detrimental change than more carefully washed material. Before about 1970, LP was usually made on a small scale and some effort was expended on freeing it from 'whey'. Since then it has, to an increasing extent, been made in bulk and has merely been pressed or, if washed, washed perfunctorily. As an example of the uncertainties thus introduced, a paper by Ueda & Ohshima (1983) may be quoted. Chicks had better weight gain and N utilisation when given LP from oats and grasses, than from beans and clover, as sole protein sources. But oven-dried curd, which had contained 65% 'whey', was used. It is as likely that the difference was caused by differences between the contaminating 'wheys' as by intrinsic differences between the LPs.

Although hazardous, heating has compensating advantages; it is a convenient method of sterilisation, and is the cheapest method for making dry products that can be easily dispensed in feeding experiments, or marketed commercially. Much more research is therefore needed on the precise nature of the changes undergone by LP from different sources, and after being pretreated in different ways, when it is heated. When this work is done, it is to be hoped that the precise conditions of washing (e.g. the % DM of the wash-water pressed from the press-cake and the % DM of the press-cake before it is dried by heating in an oven), drying and heating will be stated. Several statements about digestibility have not been quoted here because, in the absence of information on any of these matters, their significance cannot be assessed.

The process of making 'chamois' leather, in which partly dried hide is beaten with the unsaturated fats in fish oil and then heated and exposed to air to oxidise the oil, is so well known that the risk of LP becoming tanned in a similar way was obvious from the start of work on LP extraction. The process probably does not involve the formation of covalent links (Shah *et al.*, 1967; Shah, 1971). There was increasing loss of digestibility when moist LP was dried at 60, 80 and 100 °C. This was interpreted as the result of physical rather than chemical changes in the protein: the loose complex formed between protein and partly oxidised unsaturated fatty acids could be disrupted, with restoration of protein digestibility, by solvent extraction. Oxidation of the fatty acids was inhibited by antioxidants such as amla (*Emblica officinalis*) powder (Shah, 1968).

Three main types of change are relevant when LP is heated during drying: effects of heat on the protein itself, effects on the lipid, and interaction between these two components of the preparation. Buchanan (1969*a, b*) dissociated the environmental conditions and measured the extractability of the lipids, the *in vitro* digestibility, and the nutritive value for rats. During storage for a few days at 100 °C or a few weeks at 60 °C with access of air, lipid became less readily extractable and the protein less digestible by papain. In conditions in which water vapour could escape freely there was no further change on prolonged heating. Loss of digestibility was obvious in 5 hours at 100 °C when LP was heated in sealed tubes so as to retain the water (about 9%) present in air-dry protein; when only 2.5% of water was present, several weeks were needed for the same loss. There was a similar loss of digestibility in the absence of oxygen, but solvent-extracted protein did not lose digestibility even in the presence of air and moisture. By solvent extraction after moist heating in a sealed tube, i.e. with restricted access of air, digestibility returned to nearly the normal value even although prolonged heating, especially when air had access to the LP, had made much of the lipid unextractable. The importance of excluding light during prolonged storage has already been commented on (p. 81). The conclusion to be drawn from these observations is that there are advantages in removing water from LP as rapidly as possible although heating in a current of air increases the risk that part of the lipid will be peroxidised. Literature on the toxicity of peroxidised lipids is surveyed by Mead & Alfin-Slater (1966). Speed may be advantageous even if, as Mauron (1970a) suggests with other foods, the changes undergone by lipids affect the digestibility of protein more by mechanical entanglement than by complex formation.

The behaviour of LP when heated, or during prolonged storage, differs in no essential respect from the behaviour of other proteins that contain unsaturated lipids. The behaviour of unfractionated LP when heated after solvent extraction has not been thoroughly investigated: there would be little reason for the investigation because the material would already be dry and sterile. Carefully made preparations of 'cytoplasm' protein are nearly free from lipid. If they were being dried by heating, the possibilities of other reactions, e.g. between the ε-NH_2 of lysine and the β or γ-COOH of aspartic or glutamic acids, and various types of oxidation, would arise. An intermediate preparation, soya LP partly freed from 'chloroplast' protein by centrifuging the original extract at 20000 g (Betschart & Kinsella, 1974b), was digested more slowly by papain after 24 weeks in air at 27 °C.

Digestion in the rumen

It is agreed that extensive hydrolysis in the rumen diminishes the value of proteins in ruminant feeding because the resulting amino acids may be deaminated by rumen microflora. Amino acids are used more efficiently if their liberation from protein is postponed until protein reaches the intestine. Furthermore, there is evidence that soluble proteins in fodder are part of the cause of bloat – the harmful accumulation of froth in the complex stomach of ruminants. The rate of digestion of LP processed in different ways, and of the products of its fractionation, has therefore been measured both *in vitro* and *in vivo*.

RuBP is initially in the chloroplasts; it is partly liberated from them during juice extraction, and then partly coprecipitated along with them during fractionation. The amount of it in 'cytoplasm' protein is therefore variable when that fraction is prepared in bulk. In cannulated sheep and cows, the rate of hydrolysis of samples of this protein (labelled with ^{14}C) was three to ten times greater if the animals had been fed on lucerne rather than hay + concentrates. Studies on rumen fluid *in vitro* suggested that hydrolysis is preceded by adsorption to bacterial cells and that different diets favour different microflora (Nugent & Mangan, 1981). The rate of digestion of this leaf protein fraction in the rumen is intermediate between those of casein and serum albumin (Nugent *et al.*, 1983), and it is diminished by competition for proteolytic sites when two proteins are present at once. These effects cannot be attributed entirely to bacteria; rumen protozoa contain enzymes which digest this leaf protein fraction more rapidly than they digest casein (Coleman, 1983).

It is unlikely to be worthwhile extracting LP for use as feed for adult ruminants. Lawes' reasons for thinking that have already (p. 4) been quoted. Although experiments in the adult rumen may not give reliable evidence about the value of LP as feed for calves, they have some academic interest. Lu *et al.* (1981) found that the temperature and duration of heat coagulation had little effect on pepsin digestibility of LP from lucerne, but that digestion in the rumen of cannulated sheep was diminished by prolonged exposure to 90 °C. That conclusion was amplified (Lu *et al.*, 1982, 1983*a*) in a comparison of material coagulated by anaerobic fermentation and then spray-dried, by heat coagulation and not dried, and by heat coagulation and spray-dried. The different treatments did not affect digestibility or transit time through the gut, but the first treatment resulted in most, and the last in least, amino acid deamination in the rumen. By all the measuring techniques used in experiments with lactating cows, LP was less readily digested in the rumen than soya bean meal (Lu *et al.*, 1983*b*).

Digestibility in vivo *and nutritive value*

Detailed experiments on animals are an essential prelude to any attempt to introduce a novel protein into human diets; they also have direct relevance when economically important animals, such as chickens and pigs, are used. Unequivocal interpretation is, however, extremely difficult. Species differ in their array of gut enzymes, in their amino acid requirements and in their susceptibility to extraneous substances that may be present in the protein. During the phase of work on LP that I have dubbed 'Natural History' (p. 43), every extraction and every feeding experiment was, to some extent, interesting. Now, an extraction is not worth discussing unless the antecedents of the crop are fully described, or a new type of machine was used for the extraction. Similarly, the results of a feeding experiment are now of little interest unless precise information is given, not only about the species from which the LP was made, but, more significantly, about the details of the separation procedure, the method of preservation, and, if the curd was dried, the amount of 'whey' that it contained. Some curds containing only 20% DM seem to have been dried; reactions between protein and the carbohydrate and phenolic compounds in the residual 'whey' are a reasonable explanation of the poor nutritive value sometimes found. More efficient removal of such contaminants is the probable explanation of the better quality of LP made by ultrafiltration rather than heat coagulation (e.g. Ostrowski-Meissner, 1980a). Poor nutritive value is probably more often the consequence of damage during processing than of genuine differences in the character of the protein synthesised in the leaf. Nevertheless, I have often been amazed at the good results given by some products which I knew, having seen the techniques used, to be heavily contaminated or damaged.

Experiments on pigs, chicks and fish (Cowey *et al.*, 1971; Ogino *et al.*, 1978) have obvious economic importance. Experiments on rats are less relevant. We are unconcerned about their nutritional welfare, and an animal with a surface-to-volume ratio differing markedly from our own is a poor model. In a homoiothermic animal that ratio, and the environmental temperature, control the probability that a diet which meets energy needs will also meet needs for specific dietary components such as amino acids and vitamins. Rats eat 10% of their body weight per day: we eat 1%. However, there is now so much knowledge about the nutritional needs of rats that experiments on them deserve attention. The use of LP for feeding foxes and mink (Milovanov *et al.*, 1985) is mentioned only because it is so unexpected. The relevance of work on other mammals, and on insects, protozoa and bacteria, is doubtful. Some of them are experimentally convenient, but these organisms are even less likely than rats to have digestive capabilities and amino acid requirements that resemble our own. In this section, results on pigs, poultry and rats only will be considered: they do not conflict with results on other organisms.

It is unrealistic to feed an animal on a diet in which all the protein comes from one source. The merits of a protein are therefore usually assessed from experiments in which the protein that is being tested supplies 10 to 20% of the total protein; the performance of the protein under test then obviously depends on the amino acid composition of the proteins in the remainder of the diet. Performance probably also depends on the precise physiological state of the test animals, on the temperature and illumination at which they are kept, and possibly on the time of year of the test. For these and other reasons, the attempts often made to grade proteins on a numerical scale, sometimes even to three significant figures, seem misguided. It is no more possible to grade proteins in this way that it would be to grade all alloy steels on one numerical scale regardless of the uses to which they are to be put. The impression of imprecision given by vague phrases containing words such as *better*, *worse* and *nutritive value* will distress orthodox trophologists, but it is less misleading than the spurious precision suggested by a number. A rat is not a chemical reagent.

The first experiment on pigs (Barber *et al.*, 1959), which showed that they grew a little faster when a predominantly cereal diet was supplemented at two levels with LP rather than with the same amount of protein in the form of a white fish-meal, met with some scepticism. Doubts were dispelled by a more elaborate experiment (Duckworth *et al.*, 1961) in which four levels of LP were compared with three of fish-meal and one level of groundnut meal mixed with LP. Again, the ratio of live weight gain to food intake was slightly greater with LP than with fish-meal. The satisfactory result with a mixture of LP and groundnut meal was particularly interesting. Lucerne LP replaced soya bean meal satisfactorily in feed for large pigs and, in some experiments, in feed for small pigs also (Cheeke,

1974, Myer *et al.*, 1975; Carr & Pearson, 1976; Kanev *et al.*, 1976) if the LP had been carefully made. Replacement of 60% of the fish-meal, dried skim milk, or yeast with lucerne LP did not affect weight gain or final carcass yield (Skorobogatykh & Aitova, 1984). Results with early commercial products were less satisfactory (Cheeke *et al.*, 1977), the quality of commercial material is now better (Bourdon *et al.*, 1980). When made on the scale needed for feeding trials with pigs, commercial LP should, in principle, be better than LP made in laboratories because the processes of coagulation and separation are quicker with industrial-scale equipment.

Apart from a trial in which LP from water hyacinth satisfactorily replaced 25% of soya bean meal in pig diets (Alcantara & Lobos, 1981), all recent trials have used LP from lucerne. Yet photo-sensitisation (p. 50) seems to have caused no trouble. The early experiments were made with LP from cereals; they contain little chlorophyllase, and the leaf juice was heated quickly (Morrison & Pirie, 1961). But lucerne juice used in recent experiments was often kept acid for several days before use. It is not obvious why pheophorbide, the photosensitiser, was not formed as it was in the experiments of Carr & Pearson (1974). Pigs in these recent experiments were perhaps more often kept in the shade. This is a point which deserves fuller study. Pheophorbide formation can, however, be avoided by heating leaf juice quickly.

Early experiments with poultry were surveyed by Woodham (1971). In most of them, the quality of LP was assessed in conditions restricting growth by slight protein deficiency so that differences would show up clearly, and they were not all strictly comparable because of differences in the manner in which LP was dried. More recent publications deal with poultry given diets used in commercial practice, but in which part of a conventional protein source is replaced by LP. Liveweight gain was hardly affected by 25% replacement, but was diminished by 75% (Korniewicz *et al.*, 1980); there was some diminution with 60% replacement by LP from first or second cuts of lucerne, or with 20% replacement by LP from third cut (Kumprecht *et al.*, 1984). Meat quality, egg number, fertility and hemoglobin were not affected. When used as the sole source of protein for young chicks, LP from oats, grass, or clover, supplemented with arginine + methionine gave only half the growth rate and N retention given by soya (Terapuntuwat & Tasaki,

1984). Hanczakowski *et al.* (1981) recommended LP from potato haulm for broilers. Supplementation with methionine did not invariably improve growth in experiments in which LP replaced soya or fish-meal.

Knowledge about the digestibility of LP *in vivo* and about the effects on its nutritive value of different methods of processing and different forms of amino acid supplementation comes mainly from experiments on rats. There is no need to discuss here experiments showing, or purporting to show, differences between LPs made from different species. They may be real: but, as already remarked, until comparisons are made between preparations separated and dried or otherwise preserved, in precisely the same way, from crops taken at several stages of maturity, the reality of species' differences will remain uncertain. The early experiments on rats need not be surveyed again; that has been comprehensively done by Woodham (1971, 1983). Instead experiments will be considered that have a bearing on the effects of phenolics and other tanning agents on reactions that destroy specific amino acids or make them unavailable, and on complex formation with lipids.

Phenolic compounds and other tanning agents

Phenolics interfere with the extraction of protein from leaf pulps (p. 23). They inactivate enzymes so that elaborate precautions have to be taken if active preparations are to be made from some leaves (p. 61). They tend to depress the nutritive value of fodder for ruminants (McLeod, 1974) but there is evidence that a suitable amount of tannin in fodder prevents the accumulation of foam that causes bloat. Generalisations on these subjects are inadvisable because of the very large number of phenolics that are present in some leaves (Milic 1972; Thakur *et al.*, 1974) and the changes that they undergo during harvesting and processing (Pierpoint, 1971, 1983; Synge, 1975, 1976; Van Sumere *et al.*, 1975).

Although people consume many sources of tannin for pleasure (e.g. wine and tea) or of necessity, (e.g. sorghum), the main factor, besides freedom from fibre, that leads to the selection of certain species as leafy vegetables, is freedom from phenolics and other tannins. By-product leaves of these species will be used as sources of LP but there is little advantage in extracting protein from a leaf that is edible after conventional cooking (p. 44). It is likely, therefore, that LP will be made from crops containing phenolics because it is easier to get large yields from them than from vegetables – perhaps the phenolics give some protection from attack by pests (Feeny, 1969; Levin, 1971, 1976).

It was reasonable to assume that conjugation with phenolics would diminish the nutritive value of LP, and there is now good evidence that it does. The long series of papers by T. Horigome and his colleagues on the digestibility and nutritive value of casein and other proteins after combination with phenolics need not be surveyed here. These investigations started with work on the digestibility in rabbits of different varieties of clover, and fractions made from them, but most of the publications deal with artificial mixtures, and my knowledge of them depends on abstracts.

A reversible association between phenolics and peptide bonds, which leads to cross-linking between protein molecules, or between parts of one protein molecule, is usually invoked to explain the diminished digestibility of proteins containing phenolics. The evidence is reviewed by Van Sumere *et al.* (1975). It is usually assessed by comparing the amounts of N in the feces of rats fed on diets containing different proteins, or no protein. The trustworthiness of this method clearly depends on the extent to which different diets abrade the intestinal mucosa, for much fecal N is derived from that source. However, the striking effect of small additions of tannin suggests that the method is valid. Eggum & Christensen (1975) found that 1.5% of tannin diminished the digestibility of soya bean protein from 93 to 73%. After allowing for that diminution, they found that the biological value of the protein was not altered. This became less surprising when they found, by analysing the feces, that it was the inessential amino acids, glutamic acid, glycine and proline, that were rendered least available. 'Chloroplast' protein is less digestible in rats than 'cytoplasm' protein (Henry & Ford, 1965; Subba Rau *et al.*, 1969, 1972); it is reasonable to assume that the greater concentration of phenolics in the former is a partial explanation. It would be interesting to know whether this difference arises mainly because the chloroplasts contain more phenolics, e.g. plastoquinone, or because combination with phenolics makes other proteins coagulate along with the chloroplasts during the heat treatment used for fractionation.

Complex formation with the ε-NH$_2$ of lysine is the most clearly defined reaction between proteins and phenolics. This, like the comparable reaction between gossypol and the proteins of cotton seed, makes lysine unavailable to nonruminants although it still appears in acid hydrolysates. The amount of lysine bound in this way was measured by Allison (1971) and Allison *et al.* (1973) by destroying with nitrous acid the ε-NH_2s of those lysine molecules that were not complexed with phenolics, hydrolysing, and measuring the lysine that had been shielded from nitrous acid. Allison *et al.* (1973) point out that nitrous acid probably penetrates the protein molecule more completely than the larger molecule of fluorodinitrobenzene, which is often used in a somewhat similar manner, and so will react more completely with unshielded lysine. In lucerne LP which had been protected from oxidative complexing by processing in the presence of sulfite, Allison (1971) found only 6.5% of the lysine shielded from deamination and therefore probably unavailable to nonruminants; 9.5% was shielded in LP made without precautions, and 18.5% when chlorogenic acid was added at the start of the extraction. The values on 15 samples from seven species, made with no attempt to prevent oxidation or complexing, ranged from 13.3 to 38.2%; the larger values were given by 'chloroplast' fractions. The nutritive value of these samples had been measured on rats (Henry & Ford, 1965); it correlated reasonably well with the amount of lysine not shielded from deamination. There is so much lysine in LP (p. 63) that the loss in nutritive value if some of it is made unavailable by complexing with phenolics is not likely to be great. This expectation was borne out experimentally (Henry & Ford, 1965; Shurpalekar *et al.*, 1969); supplementation with lysine did not improve the nutritive value of rat's diets containing LP. Woodham (1971) quotes another trial (Reddy, unpublished) with the same result, but Bickoff *et al.* (1975) found lysine supplementation advantageous.

The measurements made by Subba Rau *et al.* (1972) of the quantity of phenolics in LP preparations have already been referred to (p. 61). One preparation, made from carrot (*Daucus carota*) tops, contained 37.2% ash and only 3.9% N: there may therefore have been other reasons besides the presence of phenolic material for the failure of rats to thrive on it. Omole *et al.* (1976) found no defect in carrot tops as a food for rabbits. The use of sulfite to prevent phenolic contamination has already been referred to (p. 62). One variety of lucerne used by Horigome (1984) contained less phenolic material or phenolase than usual: reducing agents caused less improvement in the digestibility of LP from it than from clover or ryegrass. Reducing agents do not, however, make all LP preparations alike. Even with their use, LP from clover did not have the nutritive value of LP from oats.

Sequestration of lysine does not seem to be an adequate explanation of the poor nutritive value of some LP preparations. The possibility that phenolics, which are often methylated before being excreted, increase the demand for methylating agents such as methionine, will be discussed later (p. 98). The position will become clearer when there has been more work along the lines started by Davies *et al.* (1975, 1978); they hydrogenated a protein:phenolic complex and isolated amino acid derivatives that had been stabilised by hydrogenation so that they withstood acid hydrolysis. By such means it should be possible to identify the amino acids that have combined with phenolics, and also to identify the particular phenolics, out of the mixture present in a plant such as lucerne, that react most readily with LP. Contrary to the assumption which seems to be made in some publications, all phenolics and other tanning agents are not similarly reactive.

There will be general agreement that complex formation between tanning agents is to be avoided if possible. Good evidence for efficacy will therefore be demanded before the suggestion (p. 82) that phenolics should be added to LP in order to protect β carotene in LP is adopted. Tanning agents exist preformed in some leaves. In many more, phenolics oxidise to tanning agents when leaves are pulped in the presence of air. More work is needed to identify the phase of the extraction procedure at which LP is most vulnerable. It would be difficult to exclude air during bulk extraction, but the use of sulfite is conventional in the food industry, is well known to produce pale preparations of LP and would be feasible economically. Such agents must, however, be used judiciously because they can destroy essential amino acids such as cysteine. Undoubtedly, the best policy is to work with species and varieties that are relatively free from phenolics: that would both increase the extractability of LP and avoid damage to what was extracted.

Modification of cyst(e)ine and methionine

A peculiarity of the distribution of amino acids in the bulk proteins from seeds and leaves is that, although minor, individual proteins, such as urease, a protein from the seeds of *Antiaris toxicaria* (Kotake

& Knoop, 1911), and trypsin inhibitors, are among the richest in S-amino acids, a deficiency in sulfur is almost universal. The presence of substances in LP that increase an animal's need for S-amino acids, or the exposure of LP to treatments that diminish the availability of S-amino acids, should therefore be avoided. Detoxication of various aromatic substances by conjugation with cysteine to form mercapturic acids has been know since the end of the last century and may explain the protection given by high-protein diets against some forms of poisoning, e.g. by *Senecio* (Cheeke & Garman, 1974). The gut flora of herbivores can, to a large extent, metabolise phenolics, but some are detoxicated by methylation. Mice methylate the breakdown products of lucerne tannins (Milic & Stojanovic, 1972). Several authors have made the plausible suggestion that the extra demand for methyl groups, when phenolics are present in the diet, increases the need for methionine. The subject is discussed by Eggum & Christensen (1975); they found no evidence for this effect in rats, though it seems to be real in chickens.

Phenolics may not only increase the need for cyst(e)ine and methionine, they may also make them unavailable. Methionine can react with *o*-quinone (Vithayathil & Murthy, 1972) and *p*-quinone reacts with other thioethers (Bosshard, 1972). There is no evidence for such reactions in the conditions to which LP is exposed during extraction. Quinones and cysteine (Roberts, 1959; Pierpoint, 1969) react in physiological conditions. Cysteine also reacts with plant lactones containing a :CH$_2$ group (Rodriguez *et al.*, 1977); this is unlikely to be a problem with LP because these lactones tend to cause dermatitis; farm workers are therefore likely to insist on the elimination of plants containing them from any crop. During preparation, and especially drying inadequately washed preparations, LP is probably more susceptible to damage by conjugation with carbohydrates than with phenolics. The formation of complexes partly unavailable to rats was demonstrated with free methionine (Horn *et al.*, 1968), soya bean (Mauron, 1970*b*) and casein (Pienazek *et al.*, 1975); the reaction has not been demonstrated with LP, but its occurrence is probable.

The amount of methionine in proteins is usually measured after oxidation to the sulfone (MSO$_2$); any sulfoxides (MSO) present are therefore included in the value. Attempts to measure the amount of MSO precisely are frustrated because it is partly reduced to methionine during hydrolysis (Njaa, 1962; Gjøen & Njaa, 1977).

By using an analytical method which avoided that reduction, Byers (in Pirie, 1970*a*) found that 18% of the methionine in wheat LP made with quick processing was present as MSO, and 30% in a sample made from juice coagulated 2 hours after expression. Ellinger (1978 and personal communication) concluded that that was the limit of MSO formation during normal processing of leaf juice.

Methionine (Figure 10), and its derivatives with a free—NH_2 group, are oxidised to MSO by sulfite in air (Yang, 1970). The mechanisms by which it is oxidised in proteins *in vivo* are discussed by Brot & Weissbach (1982). The stereochemistry of the oxidative process has not yet been investigated in detail. The :SO group in MSO exists in two stereoisometric forms. In nonenzymic oxidation, both should be formed equally unless the asymmetry at the α carbon atom exerts an influence. One isomer will be probably predominate if MSO is made enzymically. There is no evidence that :SO isomers differ nutritionally – but the possibility should be borne in mind. They may, for example, have different effects on enzymic hydrolysis of a peptide bond near the methionine.

The nutritional status of oxidised methionine is not completely clear. There is agreement that MSO_2 cannot replace methionine (Miller & Samuel, 1970); much of it is excreted as the *N*-acetyl derivative by rats (Smith, 1972). Conflict over the value of MSO arose mainly from work with chickens. Gjøen & Njaa (1977) found MSO fully equivalent to methionine in rat diets in which there was adequate cyst(e)ine, i.e. the well-known partial interchangeability of S-amino acids does not, in all circumstances, extend to MSO.

Figure 10. The structure of methionine, methionine sulfoxide and methionine sulfone.

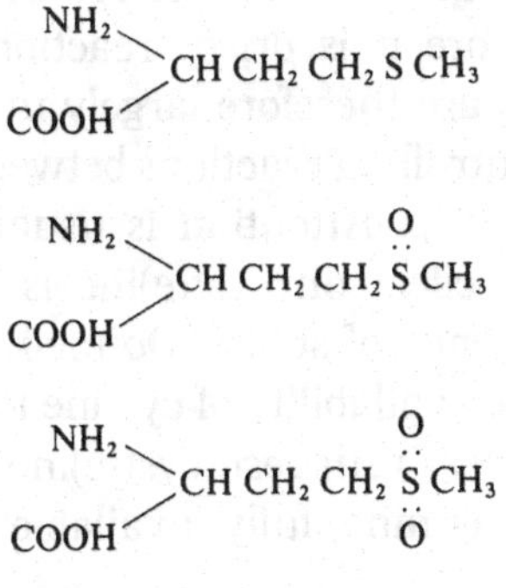

Methionine

Methionine sulfoxide

Methionine sulfone

More aspects of the subject are discussed by Njaa & Aksnes (1982). The value of MSO in human diets is unknown.

Rats grow faster on diets supplemented with LP from ryegrass or oats rather than from clover (Horigome, 1977; Horigome & Uchida, 1980). This was explained by the smaller methionine content of clover LP. The explanation is plausible, but Walker's (1982) observation that cystine and methionine measurements are liable to error should be borne in mind. Byers (1975) argued from total sulfur measurements that LP from lucerne and lupin contained 2.3 to 2.5% cyst(e)ine. That amount, if all of it is available, should be adequate.

There is therefore no evidence that the formation of MSO is the cause of the general observation (Henry & Ford, 1965; Shurpalekar *et al.*, 1969; Trigg & Topps, 1971; Hove *et al.*, 1974 and many publications since these) that the growth of rats increases when methionine is added to diets in which LP is an important component. Byers (1975) pointed out that LP usually contains more than the 3.5% of S-amino acids which is thought adequate by FAO (1973*a*) (p. 113). A partial explanation may be that people, for whom the FAO standard was drawn up, do not need as much methionine as do young rats; the rat model may undervalue LP as a human food. The situation with pigs and chickens is uncertain. Methionine supplementation was beneficial in only some experiments. Hanczakowski (1983) added another complication with the observation that sodium sulfate increased the growth rate of chickens on a diet containing lucerne LP.

In spite of much research there is still no certainty about the nature of the factor(s) causing part of the methionine and cyst(e)ine in LP to be unavailable, so that soya protein, in spite of its smaller content of S-amino acids, has usually been found a better supplement. It is a pity so little effort has been put into removing 'whey' components from the curd before it is dried: reactions between carbohydrates and amino acids are therefore largely uncontrolled. More effort has been put into controlling reactions between phenolic compounds and the S-amino acids. Attention is usually concentrated on methionine, but the effect on cyst(e)ine is even more striking. By pulping in the presence of sulfite, Donnelly & Rattray (1983) increased the percentage availability of cystine for rats from 27 to 73. If, as Gjøen & Njaa (1977) suggest, cyst(e)ine is involved in making derivatives of methionine fully available, the most important technical improvement now needed is a method for preventing damage to cyst(e)ine.

Unsaturated fatty acids

Although many foods are esteemed because of the flavour of oxidised fats, it is generally agreed that oxidation is not nutritionally beneficial and may be harmful (Mead & Alfin-Slater, 1966). The conditions in which lipids in LP oxidise were discussed in the section on preservation (p. 76); diminished digestibility associated with heating and drying has been discussed (p. 79). Buchanan (1969*a, b*) found that wheat LP, damaged by heat, could be restored to nearly its original state by thorough extraction with lipid solvents. Because fine grinding had not the same restorative effect, he dismissed the suggestion, made to explain heat damage to fish protein, that the lipids spread as an impermeable film over the heated particles of protein. This is not altogether convincing. He found that prolonged soaking restored digestibility to LP that had been extracted with acidified solvents: there is no reason to think that sufficiently thorough grinding would not have been efficacious. The point deserves attention because, although the advantages of working with fresh, moist LP have been stressed (p. 80), dry material will often be used. During the short sojourn in a rat's gut, hard dry material may not have time to soften to the extent necessary for enzyme action. LP may therefore often appear to have poor nutritive value simply because it was badly dried. On the other hand, there are innumerable ways in which oxidised lipids could react damagingly with amino acids (Gardner, 1979). That possibility should therefore be kept in mind.

Supplementation of cereal-based diets with leaf protein

The protein in interesting and palatable diets is, and should be, derived from several sources. There is therefore a lack of realism in experiments in which a concentrate such as LP is used as the sole protein source. When proteins from several sources are being eaten regularly, it is the distribution of amino acids in the whole diet that is important: an essential amino acid deficiency in one component may be compensated by an abundance of that amino acid in another component of the diet. The value of LP as a supplement to barley in pig feeding has already been referred to (p. 92). All experiments quoted here were on rats.

LP supplemented a rice diet (Sur, 1961), and a stimulated Ghanaian diet in which rice was an important component (Miller, 1965). In Miller's experiment it was a marginally better supplement than skim milk. That unexpected result may have been caused by β carotene in the LP. An unpublished experiment by Ghosh (in an annual report, and in Pirie, 1970*b*) deserves mention because LP from water hyacinth, a troublesome weed, supplemented rice when the LP supplied 30, 40 and 50% of the protein in the diet; its value diminished when more was used. The effect of varying the ratio in which two sources of protein contribute to the total in a diet is clearly shown in experiments by Subba Rau & Singh (1971). In two experiments, the weight gain was greater when half the protein came from wheat grain and half from lucerne LP than with any other ratio. It was argued that this is what would be expected from the amino acid composition of the diets: at that ratio the relative abundance of cyst(e)ine and methionine in wheat compensates for the apparent scarcity of the S-amino acids (p. 100) in LP, and the relative abundance of isoleucine, lysine, threonine and valine in LP compensates for their scarcity in wheat-grain protein. Supplementation of wheat grain by LP from three species was also observed by Garcha *et al.* (1971) and Kawatra *et al.* (1974).

Possibly deleterious components of leaf protein

Many species which are widely used as forage crops are, to varying extents, poisonous. The substances responsible tend to be more harmful to nonruminants than to ruminants. Therefore, if they are carried through into LP, they may harm poultry, pigs and people. Most of those which are soluble in water at pH 4 should be removed in the 'whey'; attention here is restricted to substances which may remain in LP which has been thoroughly pressed and washed.

LP from a few species, e.g. maize and peas, can be used unwashed because it has a pleasant flavour if used soon after preparation and before there have been Maillard and similar reactions. Although most of the estrogenic isoflavones in red clover are removed in the 'whey', Glencross *et al.* (1972) found four adhering to the LP. Two of these isoflavones, biochanin A and formononetin (Figure 11), were present in quantities which were together equivalent to the presence of 3 μg of diethylstilbestrol in 100 g of LP. If LP from species carrying these isoflavones were being used regularly, it could become necessary to devise methods for removing or destroying these substances. On the other hand there are claims, based on experiments with rats, that some of them diminish the amount of cholesterol in blood and so could prevent atherosclerosis. More evidence for and against them is needed.

Although young lucerne is sometimes used as a green vegetable after cooking in the normal manner, the presence of harmful substances in it is well known, e.g. Ferrando & Spais (1966). Several examples have already been referred to (pp. 22, 46, 50, 54, 60). An extreme example was the death of chicks given lucerne LP as sole protein source (Ueda & Ohshima, 1983). When used to supply only part of the protein, no problems arose with pigs (Prokop *et al.*, 1984) or chicks (Ameenuddin *et al.*, 1983, 1984*b*; Kumprecht *et al.*, 1984), especially when the varieties relatively free from saponins were used (Ameenuddin *et al.*, 1984*a*). Unfortunately, there has been a trend recently towards the cultivation of saponin-rich varieties because of their supposed greater winter hardiness and resistance to pests.

After washing at pH 8.5, lucerne LP contains no more coumestrol (another estrogen) than is usually present in Brussels sprouts (*Brassica oleracea*) or peas (Knuckles *et al.*, 1976). An unidentified growth depressant was removed by an acid wash (Bickoff *et al.*, 1975). In spite of defects, lucerne will probably remain an important source of LP: it should never be used as the main source of protein.

Figure 11. The structure of coumestrol, formononetin and biochanin A.

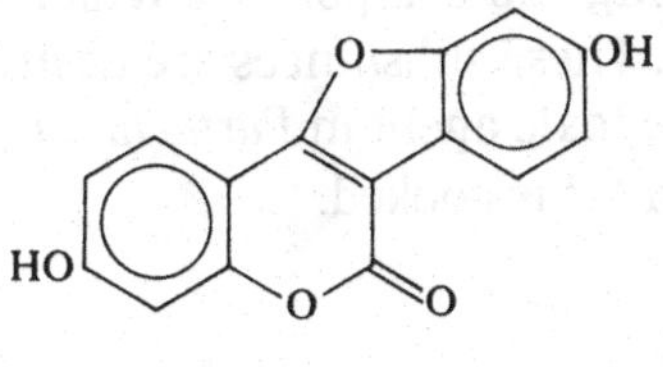

Coumestrol

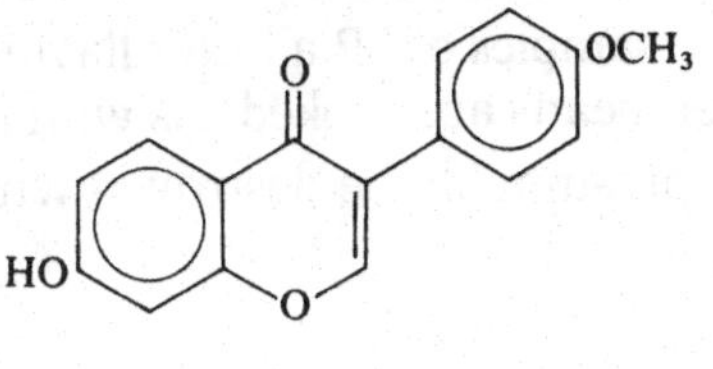

Formononetin

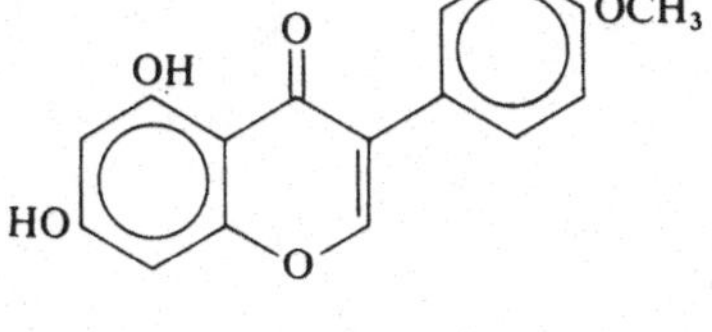

Biochanin A

More work is needed on selecting less toxic varieties, and on techniques for purifying LP from it.

The position with potato leaves is somewhat similar. They contain a mixture of slightly toxic alkaloids, often collectively called solanin. These are soluble in weak acids and could, if necessary, be almost completely removed by thorough washing. Samples of potato LP separated and washed in the usual manner contained 0.05% solanin. Potato tubers sometimes contain 0.02% calculated on the fresh weight i.e. 0.1% in the DM. The legal limit is less than that, but it is under review. No one suggests that LP should be eaten on as great a scale as potatoes, so even casually washed material should be harmless: preparative technique can probably be improved so as to meet any agreed legal limit.

Parthenium hysterophorus has become a troublesome weed in India. Savangikar & Joshi (1978) extracted LP from it. The raw weed causes rhinitis and contact dermatitis but the responsible agent(s) seem to be destroyed by heat. Ramappa *et al.* (1986) used the heated leaf meal successfully in poultry diets for broilers and layers. Presumably, heat would have the same effect on *Parthenium* LP.

Many legume seeds contain hemolytic substances which are classified as saponins: there are similar substances in leaves. The amount in LP from quinoa and *Atriplex caudatus* and *hortensis* increased as the plants matured (Carlsson, 1975). There is no reason to think that the amount present in LP, if eaten at the usual level, is harmful, but they are among the components which give some samples of LP a bitter flavour. These substances are destroyed when beans are cooked. As with the toxic agent in *Parthenium*, they will presumably be destroyed when LP is cooked.

7

Human trials and experiments

As soon as LP of reasonable quality was being made, the press-cake was regularly eaten as it was being removed from the filter stockings. We had by that time become accustomed to handling green material, and found the product from most species palatable. By 1957 we regularly cooked LP and during the next few years, several cooks were employed for short periods so as to get as many new ideas on presentation as possible. For reasons that have already been stated (p. 80), green, moist press-cake was almost always used. The initial disquiet that every cook felt at the texture and appearance of LP disappeared after one or two weeks' experience with it. That interval has an interesting historical counterpart. Woodham-Smith (1962) records that during the famine caused in Ireland by inept handling of the consequences of potato blight, corn meal from the USA was introduced on 19 March 1846. Distaste was strong enough to cause riots. By 30 March it had become 'immensely popular' and there were near-riots when the supply failed. Other examples of the falsity of the assumption that food habits are fixed are discussed elsewhere (Pirie, 1972, 1975b, 1984c).

Within broad limits, the flavour that is acceptable in food is a matter of convention and familiarity. People accustomed to bland foods accept the more strongly flavoured dishes of South India or West Africa slowly; many never accept the somewhat putrefactive flavours of some cheeses and fermented foods. Food with little or no flavour may be considered dull, but it is seldom considered inedible. Fortunately, properly washed LP from many species, e.g. cereals, cowpea and pea haulm, is nearly tasteless when fresh. A slightly fishy smell, similar to that of China tea, develops as it ages; this is presumably because of the breakdown of choline. Later, oxidation products give dried material stronger flavours. Flavour

105

adheres more strongly to even fresh LP from the clovers, lucerne and potato haulm. It is a pity, therefore, that lucerne is so often used initially – it imposes an extra barrier to ready acceptance by discriminating adults. However, this may be advantageous when LP is being made primarily as a food for children. They are more likely to get the food if it lacks a strong appeal for adults. Nevertheless, when comments are made on the flavour of LP, the species from which it was made should be stated, and the competence of those who made it should be considered. More work is needed on species from which bland LP can be made.

Presentation of leaf protein in acceptable forms

Fresh LP, and LP preserved with salt, sugar or vinegar, blends smoothly into many foods; after drying or prolonged freezing it may be gritty and must be ground thoroughly before use. Freeze-dried material has a smooth, soft texture, but that method of preservation is too expensive to be used as a routine technique.

Some forms of presentation are obviously unsuitable. Every type of baked product should be avoided because baking enhances the flavours of breakdown products of lipids; furthermore, the familiar appearance of bread has so many associations that a greenish version of it is unlikely to be acceptable. LP blends well with savoury flavours, or fish, and less well with coffee, orange or lemon, but banana is an excellent flavouring agent. A mixture of LP and banana pulp has a texture that is well adapted for spreading or insertion into some type of casing.

Little would be gained by devising dishes which contain only 1 or 2 g of LP in a helping: that amount can as conveniently be eaten as DGLV. On the other hand, LP is intended as a supplement, and it is for many reasons desirable that the protein in a diet should come from many sources. A reasonable helping of the supplemented food should therefore contain 6 to 10 g: a fifth or tenth of the total desirable daily supply of protein. Nevertheless, some of us have eaten 30 g daily for several periods of a week or more; the unabsorbed breakdown products of chlorophyll coloured feces green, but we noticed no digestive disturbance. As already mentioned (p. 10), LP was eaten by a distinguished party in 1940.

The problem of presentation can be considered under three distinct heads. When LP is being given to motivated people, or to people who have accepted the idea that it is a reasonable component of a regular diet, the simplest method of presentation is as good as any other (Morrison & Pirie, 1960). Crumbled moist protein was sprinkled at the table onto a risotto or some similar fairly highly flavoured but protein-deficient dish. The unblended particles of protein are by no means unattractive so long as one has added them oneself; it is more difficult to get acceptance if the mixture is made in the kitchen. In some dishes the protein was made into pieces, encased in thin batter or pastry, that can be eaten without being bitten and examined, or that are bitten only once. For obscure reasons people seem to be less concerned about the internal appearance of small pieces of food than they are about large ones. This can be readily confirmed at cocktail parties or in an Indian or Vietnamese restaurant. It would be interesting, and perhaps useful, to try to make simulated currants that could be put into baked products, and from which the colour would not diffuse out. In that moist form the LP is nearly black. We assumed that the appearance that people accept as normal in a food is purely a matter of convention and that most people would accept the somewhat novel appearance of our products after they became familiar with them. Similarly, people habituated to LP accept its flavour so that a larger proportion can be added to a food.

Problems arising under the second head are, to some extent, unreal. Before a novel food can be tried in any country, those responsible for agricultural and food policy, and representatives of international agencies and other grant-giving bodies, have to be given the food in an attractive form that can be eaten casually during a visit to the laboratory. We therefore devised a set of variously decorated and flavoured morsels, each containing 2 to 3 g of LP. There is no need to describe them here because they have no role in practice, but until such morsels have excited interest, nothing can be done to meet the simpler requirements of the needy. We have not been invariably successful; but only one visitor, unfortunately a senior British civil servant, has made the inane comment 'I prefer beef-steak'.

Work on the third type of presentation was undertaken at a time when most experts considered protein deficiency commoner than energy deficiency. Various dishes were therefore described (Byers, *et al.*, 1965) that could be used as protein supplements to accompany diets containing adequate amounts of the other necessary components. Arbitrary standards decided on for these supplements were: 25% of the DM should be protein, and more than 50% of the protein should be LP. Total N was measured on a sample of every suitable dish after it had been freeze-dried and ground. The weight of N contributed by each component in the dish was either determined by analysis, or taken from food tables, or both. These figures enable an estimate to be made both of the final N content to be expected and also the proportion of that N present as LP. We did not accept our estimate of the contribution made by LP unless the observed percentage of N agreed reasonably well with the estimated percentage.

Ten mixtures and methods of cooking were described (Byers *et al.*, 1965) that satisfied these criteria. Two may be given as examples (Tables 3 and 4).

Curry cubes. Flour was mixed with a little water to a cream, and boiling water, in which a bag of mixed herbs had been boiling for half an hour, was added slowly to the cream. This was cooked to make a smooth viscous sauce. Curry powder was added to fried chopped onion, and, after further frying, was added to the sauce. Finally sodium monoglutamate and LP were added. After thorough mixing the paste was frozen in trays 1 cm deep and freeze-dried.

Table 3. *Curry cubes*

Contents	Weight taken (g)	Dry weight (g)	Nitrogen content (g)
Flour	28.0	24.40	0.479
Onion	24.0	1.73	0.036
Curry powder	10.0	10.00	0.152
Margarine	20.0	17.25	0.006
Sodium monoglutamate	4.5	4.50	0.336
Barley leaf protein (freeze-dried)	20.0	19.10	2.040
Water	276.0	—	—
	382.5	76.98	3.049

Nitrogen as a percentage of dry matter:
   Calculated   = 3.96
   By analysis   = 4.23
The percentage of nitrogen due to leaf protein = 67.0

The cooked flour holds the mass together so that it can be cut into cubes; they keep at room temperature for several weeks and indefinitely in the refrigerator. This mixture is designed for people who like a fairly strong curry. If less flour is used the cubes are more fragile; this does not matter if they are being kept for visitors to sample in the laboratory, but if they are being made for demonstration elsewhere the full quantity of flour should be used.

Banana and leaf protein pie. LP was added to mashed banana and mixed well with water. The mixture was put into a pastry-lined tin and covered with a pastry lid; the pie was cooked for 25 min. It can be served hot or cold. A sweeter tasting mixture can be made by adding sugar to the pie filling.

In Nigeria and other parts of Africa, ground, dried leaves are often used as a relish; dark green material is therefore not considered unusual. Methods similar to those described above were used successfully by Oke (1966, and in Pirie, 1971a). Akinrele (1963), though he was uncertain about the economics of producing LP in Nigeria, had no doubts about its acceptability. Dark green vegetables are no longer esteemed foods in India and Pakistan; when large amounts of LP are added to familiar foods in these countries, the usual comment from adults is that the foods taste 'leafy'. In two forms of presentation, LP from cotton was judged better than LP from *Gliricidia* (Balasundaram & Samuel, 1968). No differences

Table 4. *Banana and leaf protein pie*

Contents	Weight taken (g)	Dry weight (g)	Nitrogen content (g)
Banana	150.0	44.0	0.270
Water	150.0	—	—
Short pastry	131.5	95.8	1.262
Mustard leaf protein (freeze-dried)	50.0	48.5	5.580
	481.5	188.3	7.112

Nitrogen as a percentage of dry matter:
  Calculated   = 3.78
  By analysis   = 3.86
The percentage of nitrogen due to leaf protein = 78.5

were commented on when 5% of LP from three species was added to biscuits (Garcha *et al.*, 1971). Experience in the USA (Anonymous, 1974; Betschart, 1977) was essentially similar. Biscuits are an expensive food, and baking modifies the lipids in LP in a manner that intensifies their flavour; this method of presentation is not likely to contribute significantly to the day's ration. Kamalanathan & Devadas (1971) tried several methods of presentation on taste panels in Coimbatore. The two most successful were 'dhal balls' and 'leaves chutney'. The components of the former were 20 g of LP containing 70% water (i.e. 4 g of actual protein), 22.5 g of redgram dhal, 1 g each of coriander leaves, curry leaves and green chillies, 8 g of onion, and 0.5 g of salt. The chutney, or Veppilaikatti, contained 20 g of moist LP, 30 g of bitter lemon leaves, 15 g of green chillies, 25 g of tamarind, 4 g of coriander seed, 5 g of salt, 10 g of oil and 1 g of mustard. Initially the taste panels did not like mixtures containing more than 20 g of LP but, with experience, more was accepted. In Pakistan, Toosy & Shah (1974) had similar success. Shah *et al.* (1981) made a more elaborate comparison with a team of experienced judges; it is set out in Table 5.

In these trials, adults were encouraged to express opinions on the character of the food offered. In such circumstances, many people

Table 5. *Assessment by an experienced tasting panel in Pakistan of the acceptability of familiar foods fortified with green, unfractionated LP made from mixed grasses and/or berseem. Each helping contained 8.6 g of added protein*

Products	Appearance	Texture	Taste	Flavour	Total score
Bread potato sandwich	9.0	9.0	6.0	6.8	30.8
Dahi Bhalle	10.0	9.0	9.8	9.0	37.8
Halwa	6.8	8.0	6.5	6.2	27.5
Kachori	9.0	8.5	7.3	7.3	32.1
Laddu	7.0	6.8	7.3	7.3	28.4
Missi Roti	6.5	6.5	7.8	6.5	27.3
Murmara	8.8	8.8	8.3	8.0	33.9
Pakoras	8.6	9.0	8.4	8.0	34.0
Peanut Maroonda	7.5	7.0	7.3	7.2	29.0
Wheat Maroonda	6.3	6.8	6.3	6.0	25.4

Note: In the scoring system used: < 2 = unacceptable; 2 to 4 = could be used in a crisis; 5 to 6 = acceptable but not suitable for use at a party; 7 to 8 = could be given to guests; 8 to 10 = not differentiated from the unfortified food.

tend to be more critical than they would have been if the food had simply appeared at a meal without comment on its novelty. Some features which cause adverse comment from a tasting panel might even be regarded as adding piquancy to the dish! These trials show that, when a familiar food has LP added to it and is then judged by a panel, the extent of acceptable divergence from the familiar standard is small. This means that extra care must be taken to ensure that the LP has been adequately washed and pressed, and that it does not have a gritty texture. These points may have been attended to more closely by Byers *et al.* (1965) than by some others who have worked on presentation; that would explain the differing degrees of success achieved. Another important factor is the duration of the trial. Many people react against every novel food on first contact but accept it when it has become familiar – a phenomenon sometimes labelled 'the first day syndrome'. Most of those eating the dishes prepared at Rothamsted were involved in various aspects of work on LP and had therefore already gained familiarity. This would also be the state of affairs if LP was being produced in a village. I have discussed the interrelations between factors such as these elsewhere (Pirie, 1972, 1982, 1983).

Experience with tasting panels and casual visitors shows that unfractionated LP, if sensibly presented, is acceptable. Nevertheless, people who should by now know better, e.g. several contributors in Telek & Graham (1983), and people with little knowledge of the subject, e.g. James & Larbey (1976), still assert that it is not acceptable. It is assumed that it is essential to make something which could oust some existing product from the shelves of a supermarket; this is also assumed by Wang & Kinsella (1976 *a, b*). Elaborate techniques are therefore proposed for fractionating leaf juice by acid precipitation or ultrafiltration rather than by heat coagulation (Miller *et al.*, 1975). The presence of the novel material in a food might then not be noticed or, if noticed because of mention on the label, might be accepted with resignation. This approach does not exploit the basic merit of LP – that it is a human food which can be made by simple processes from local crops for local use. People in poor countries could not afford the bland, colourless, soluble product: people in rich countries do not need it. The only exception to that generalisation is that the fractionated material, especially the crystalline fraction which can be made from tobacco (Tso & Kung, 1983) and which retains solubility (Sheen & Sheen,

1985), could have pharmaceutical uses, e.g. for parenteral feeding. Long experience with plant viruses shows that the proteins in juice from uninfected leaves are relatively poor antigens.

Measurements of the nutritive value of leaf protein

Unless adults are as strongly motivated as the conscientious objectors studied by Hume & Krebs (1949), an experiment on them would not be trustworthy unless they were confined in a penal or refugee institution. Use in such circumstances would give a novel food an association with misfortune: a stigma, once acquired, is hard to erase. Furthermore, except in prolonged trials in which epidemiological or work-pattern results are significant, it is hard to find a valid quantitative index of the merits of a food. Adult protein need is sometimes assessed by measuring 'N balance' – the amount of N in the food eaten during several days is compared with the amount excreted. It is obvious that the food contains too little protein if the amount of N excreted exceeds the amount eaten. But there is no reason to assume that the protein intake is optimal as soon as enough is being eaten for balance to be struck. It is just as reasonable to assume that some excess of circulating amino acids is beneficial (p. viii).

With children, growth rate is a good index for the merits of a protein supplement; they are the group now most likely to be malnourished, they do not have such well-established food prejudices as adults, and those living in institutions get controlled diets which are often meagre or even inadequate. Valid measurements of the effects of supplementing the diets of some of these children can therefore be easily made. It is more difficult to assess the effects of supplements on children living at home but, as will be shown, it is possible.

For obscure ideological reasons, some people condemn research on children as unethical. What has been done in these feeding trials differs in no ethically relevant respect from what is done when child mortality is compared between two comparable communities, only one of which has had the presumed benefit of a new sewage system or prophylactic vaccination. If the innovation proves beneficial, all will get it as soon as it can be afforded. The children in these trials were already malnourished and protein may have been one of the dietary components needed. Amino acid analyses given in Table 6, as well as those already given (p. 63), show that LP satisfies the specifications (FAO, 1973a) suggested for a protein supplement suitable for children. Even if it were assumed, on the basis of animal experiments, that LP is so nearly ideal that it should be given to all malnourished children, that would not be feasible because supplies are at present inadequate. In these circumstances it is difficult to see how an ethical issue arises when the effects of LP supplementation are observed.

Another group of critics of human trials of LP argues that there is no need for outsiders to promote a novel form of food because, if the world's resources were properly exploited and distributed, there would be enough conventional food for all. That is half true. The foods on which the poor in wealthy countries depend could be produced and distributed so as to meet all needs. To many of the recipients in poor countries, these foods would be just as novel as LP: food donated by outsiders under Aid Programs is already in that category. However, the basic flaws in the point of view of this group of critics are: the hoped-for social revolution will take decades to achieve, whereas the malnourished are with us now; leaf protein in some form is likely to feature in human and animal diets as farming becomes more efficient; the management of LP production and use is already coming under local control.

Criticism of the second type will continue indefinitely; it is an automatic reaction, sometimes valid, sometimes not, to every form

Table 6. *Amino acid composition of leaf protein compared to the composition suggested by FAO (1973a). (Weight of amino acid (g) in 100 g of protein)*

Amino acid	Leaf protein	FAO
Isoleucine	4.4 to 5.7	3.7
Leucine	8.4 to 10.7	5.6
Lysine	4.8 to 7.3	7.5
Cystine + methionine	2.7 to 5.2	3.4
Phenylalanine + tyrosine	10.0 to 12.7	3.4
Threonine	4.8 to 5.4	4.4
Tryptophan	2.0 to 3.0	0.46
Valine	5.6 to 6.5	4.1

Note: These are reasonable values for the desirable amounts of the essential amino acids in a protein. No similar recommendation is included in *Energy and protein requirements*: the report of a FAO/WHO/UNU committee published by WHO in 1985.

of innovation. Criticism of the first type, valid or not, has become unnecessary. There is no further need for experiments of the type described in the next few paragraphs. The point has been made: a supplement of LP improved the growth and health of many children living on inadequate diets.

Quantitative experiments on children (Waterlow, 1962) started 10 years after LP had been eaten regularly by those working with it in Rothamsted. One infant gained 1.2 kg in the 20 days during which half his protein was LP, but the main experimental periods were too short for measurements of growth to be valid; instead, the percentage of the dietary N that was retained (i.e. that did not appear in urine and feces) when the N was supplied mainly by milk, was compared with retention when part of the N was supplied by milk and half to three-quarters by LP made at Rothamsted. Table 7 summarises the results on ten infants, aged 6 to 20 months, who were part of a much larger group coming into hospital in Jamaica in an acutely malnourished state; their mean weights were only 56% of the USA standard. It is clear from the table that at the greater intake level (765 and 776 mg N daily kg^{-1}), N was not so well retained from the milk+LP mixture as from milk alone. When the amount of N in the diet was smaller, though still as great as in most feeding schedules, retention on the two diets was approximately equal. That result agreed with expectation: LP is not as good as milk, but when the supply of milk is inadequate, supplementing the milk with LP is better than relying on a small amount of milk alone. This is shown by the column on the extreme right of Table 7 which shows that little N was retained when the infants were given approximately the same amount of milk as was in the mixture used in the column adjacent to it. LP is not as digestible as the proteins in milk, and an infant develops an adequate set of gastro-intestinal proteases slowly. It can therefore be argued that when infants are still getting some maternal milk, it would be better to let the mother rather than the infant have an LP supplement.

There were no problems with acceptance, and no instances of gastro-intestinal disturbance, but two infants developed an erythema with some swelling of the feet and face. This responded to anti-histamine drugs and was thought to be an allergy. Nothing similar has appeared in later experiments. The most reasonable explanation is that our standards of hygiene were not as high at that early date as they were later.

Growth was the primary criterion used in a more prolonged trial in an institution near Mysore (Doraiswamy *et al.*, 1969). The normal diet contained ragi (*Eleusine coracana*) flour, beans, vegetables, skim-milk powder, oil and sugar. It supplied 39 g of protein day^{-1} and 7 MJ. Eighty boys, 6 to 12 years old, were divided into four matched groups. One group remained on the institutional diet; one had a daily supplement of 0.5 g of lysine (lysine deficiency is the principal shortcoming of ragi); one group had 24 g of low-fat sesame (*Sesamum indicum*) flour; and one group 15 g of lucerne LP. The last two supplements contributed 10 g of protein to the daily diet and sugar was withdrawn from all supplemented diets to compensate for the energy in the supplement. The results after six months are set out in Table 8. It is reasonable to assume that a group getting milk, if it had been included, would have responded even better than the group getting LP. That control would, however, have been unrealistic in a country where milk is scarce. N balance was measured on some of the boys three months after the experiment started. Digestibility on the control and LP diets were 55 and 66%, respectively, and the N balances (i.e. the amounts of N

Table 7. *Mean values of absorption and retention of nitrogen by malnourished infants at different levels of protein intake from milk (M), or leaf protein and milk mixture (LP + M)*

Diet	M	LP + M	M	LP + M	M
Number of infants	11	5	10	5	5
Weight gain (g per kg day)	5.4	5.2	3.6	3.4	3.2
Energy intake (MJ per kg day)	0.66	0.61	0.53	0.59	0.57
Nitrogen intake (mg per kg day)	776	765	504	496	238
Nitrogen absorbed (mg per kg day)	690	602	452	436	188
Percentage of nitrogen absorbed	89	79	90	88	79
Nitrogen retained (mg per kg day)	276	246	165	160	45
Nitrogen retained as a percentage of intake	36	32	33	32	19
Nitrogen retained as a percentage of nitrogen absorbed	40	41	36	37	57

retained by an individual daily) were 0.76 g and 1.89 g. The values on the other two diets lay between these values. Taking the groups in the same sequence as in Table 8, and judging by mainly visual criteria, the numbers showing improved nutritional status at the end of the trial were 3, 11, 8 and 13. It is clear therefore that LP is a useful supplement. The boys ate all the diets readily.

In a less fully controlled trial in Nigeria (Olatunbosun, *et al.*, 1972; Olatunbosun, 1976) the mothers of 26 children, two to six years old and suffering from kwashiorkor, were supplied with powdered dry LP made from maize and other crops. About 10 g of LP was mixed at home with the usual food of each child. After five weeks, the average total serum proteins increased from 4.9 to 6.6 g 100 ml^{-1}. Within 10 days peripheral oedema disappeared, appetite improved, there was less diarrhoea and an increase in mental alertness. Because of the disappearance of oedema, there was at first no increase in body weight, but it increased towards the end of the trial. The authors (in a personal communication) attach particular importance to the prompt recovery from the dejected apathy that is usual in kwashiorkor; the photographs in their paper give striking confirmation of this. Obviously, it is possible that the mothers surreptitiously gave these children other supplements; they were not continuously in hospital. This seems extremely unlikely because, if other supplements had been available, the children would not have got into the original malnourished state.

Because of the success of this trial, pilot-plant production was started using *Celosia argentea*. This is known locally as 'soko'; the

Table 8. *Mean measurements on four groups of boys, six to twelve years old, given different diets for six months (from Doraiswamy et al., 1969)*

Dietary supplement	Increase Height (cm)	Increase Weight (kg)	Increase in hemoglobin (g per 100 ml)	Red cell count (millions per mm^3)
None (control)	2.2	0.47	0.29	0.06
0.5 g lysine	4.25	1.05	0.64	0.22
10 g protein in sesame flour	3.51	0.86	0.73	0.19
10 g protein in LP	4.84	1.28	0.87	0.23

LP is therefore called 'sokotein'. It was given regularly to malnourished children in hospital in Ibadan. *Amaranthus caudatus*, cassava and cowpea are being tried because LP from them has a more attractive colour. The Institute of Church and Society in Ibadan set up a unit to produce enough LP to supply 10 g daily to 200 people.

Using some of the dishes listed in Table 5 (p. 110), Shah *et al.* (1981) and Shah (1983) compared the effects of supplementing the normal diets of children 7 to 14 years old with 200 g of milk or 8.6 g of protein in the form of LP. For the first month, attendance by those getting milk was 3% better than by those getting LP; after 7 months, attendance by those getting LP was 9% better. After 8 months, the control, milk, and LP groups gained respectively 1.08, 2.45 and 2.62 kg and grew 2.64, 4.87 and 5.33 cm. Changes in hemoglobin were not so spectacular but still gave LP a slight advantage. By most criteria, LP is not nutritionally as good as the proteins in milk, these results are therefore somewhat surprising: the most obvious explanation is that the β carotene in LP (p. 123) supplemented diets otherwise slightly deficient in sources of vitamin A.

Since 1975 an elaborate feeding trial has been proceeding in South India. Dr R. P. Devadas, director of the Sri Avinashilingam Home Science College, Coimbatore, became interested in 1966 in the use of LP as a human food, and organised in the college the meeting that resulted in the IBP handbook on LP (Pirie, 1971a). When Mrs C. Martin, the Chairman of the charitable organisation 'Find Your Feet', had collected enough money from various sources to finance a proper trial, Coimbatore was the obvious centre for the trial. Because of the very limited amount of control that can be exercised over what people eat, it was clearly impossible to label a properly randomised set of children and feed to each a known diet. Instead, six villages were chosen that are near Coimbatore and as similar as possible in character, type of employment and average income. A nursery school (balwadi) was set up in each village, and in this, on six days a week, about 60 children, 24 to 54 months old, spend most of the day. On the six school days they get two meals which contain 80 to 90% of their daily energy and protein supply. For various reasons, mothers were at first reluctant to send children to the balwadis, but once a nucleus of attenders was seen to thrive, recruitment was no problem. Other neighbouring villages are envious of the chosen six. The balwadis have a record of educational and hygienic success that would justify them even if they had no nutritional purpose.

One balwadi is primarily educational and the food served in it is modelled, in quantity, character and quality, on food usually eaten at home. In five balwadis the food served contains 1.3 MJ more energy than the food served in the control balwadi. In one of these five, most of that energy comes from tapioca which supplies only 1 g of protein. The food given in the other four balwadis contains about 10 g of extra protein given in the form of horse gram, maize + Bengal gram, skim-milk powder or lucerne LP. The energy contents of the supplements are equalised by suitable adjustment of tapioca and sugar (jaggery). So much supplementary energy is being given that the problems tackled by Byers *et al.* (1965) (p. 107) do not arise. The mixture used for the daily helping contains 18 g of unfractionated dried lucerne LP, i.e. 10 g of actual protein, 40 g tapioca, 10 g ragi, 2 g sesame and 30 g jaggery. The moist balls into which it is shaped resemble a sweetmeat (laddu) that is often eaten, and the molasses flavour of jaggery overpowers the taste of lucerne. There have been no medical problems, and the laddus are so well liked that children who are not among the 60 included in the trial gather round to get any surplus remaining after the 60 have been supplied.

In a trial such as this the results cannot be assessed as easily as in a trial, such as the one in Pakistan, with a fixed group of participants. Children leave when they reach the age at which they go to elementary schools, and are replaced by younger children. Nevertheless, some clear results had emerged after 18 months: they are set out in Table 9. That Table includes measurements of hemoglobin in those children who were willing to be pricked for a blood sample. The general clinical picture was assessed from time to time. As a result of education and attention, there was improvement in all the balwadis. It was particularly striking in the one getting LP.

It is clear from this trial that milk, as would be expected, is a better supplement than LP, but LP is marginally better than the other two sources of protein. Little emphasis should be put on that point. The object of the trial was not to arrange the locally available protein sources in order of merit, but to see whether LP should be classified among the acceptable and useful local sources. It is clearly established that it should be so classified. Apart from establishing the

Table 9. *Results of an 18-month comparison of different supplements to the diets of pre-school children*

| | | | Diets supplemented with 1.3 MJ | | | | |
| | | | Diets supplemented with 10 g protein | | | | |
	Age (months) at the start	No supplement	Tapioca	Horse gram	Maize + Bengal gram	Milk	Leaf protein
Increased height (cm)	24–30	9.6	10.0	10.3	10.8	12.0	10.6
	36–42	9.3	9.4	9.8	10.3	11.8	10.3
	48–54	8.8	8.3	8.9	10.1	10.8	10.1
Increased weight (kg)	24–30	2.7	2.8	3.0	3.0	3.4	3.2
	36–42	3.0	3.1	2.9	3.3	3.6	3.2
	48–54	2.9	2.9	3.0	3.0	3.4	3.0
Increased hemoglobin (g in 100 ml)	24–30	2.3	2.1	2.3	1.6	2.2	2.5
	36–42	2.4	2.9	2.4	3.1	3.2	3.5
	48–54	2.4	2.5	3.0	3.7	4.3	4.2

value of LP, the main points that arise from the trial are that children readily accept LP laddus, and their mothers are unperturbed by the greenish tinge of the feces. Another point worth noting is that the group getting extra energy without extra protein shows little improvement over the control: this tends to contradict the statement, now often made, that lack of energy is more common, in regions such as South India, than lack of protein. A subsidiary trial (Devadas *et al.*, 1978) showed that half of the extra N supplied by 7.5 g of protein, in the form of LP, was retained.

Trials continue in Coimbatore along similar lines. The main difference is that there are now only five groups: milk, LP, LP with an additional 1.3 MJ given again as tapioca so as to make a 2.6 MJ supplement, 1.3 MJ supplement, and control. The results essentially confirm those in the earlier trial (Devadas, 1981; Devadas *et al.*, 1984). Milk is still the best supplement, but the extra energy is beneficial. This suggests that the amount of energy supplied by the food eaten at home was not as large as had been thought. Uncertainties such as that cannot be avoided in experiments on unrestrained people.

As a result of these experiments, which establish the position of LP among the protein concentrates which could beneficially be included in human diets, attention is now turning to trials aimed at finding out how its production and use can be fitted into the normal life of a community. In Saltillo, a village 65 km south-east of Mexicali (Mexico), 'Find Your Feet', cooperating with Professor H. D. Bruhn, started a trial in 1983. Pulp is made with an electric mincer (Bruhn *et al.*, 1983) and juice is extracted in a hand-operated press. After heat coagulation, the moist green press-cake is used within a few days of being made. Of several methods of presentation, the simplest seems to be the most successful: spreading along with highly seasoned red bean paste on tortillas. No problems have been encountered in getting 2- to 14-year-old children to take 15 g DM of LP daily, i.e. about 8 g of actual supplementary protein. The initial demonstration on 58 children was so successful that 100 are now involved and there are plans to start similar projects, run by 'Mothers' Cooperatives', in four more villages. Because of hot, dry weather in summer, production is not continuous and only the results of 4.5 months of supplementation have been analysed. Beneficial results are however clear among children less than 6 years old, and among those who were initially malnourished. A larger unit, managed on a commercial basis by a cooperative, was established in 1984.

A unit at Bidkin, near Aurangabad (India) was originally planned to be self-contained (Joshi *et al.*, 1983, 1984); the crop was to be grown and fractionated on a farm, the fibre fed to cattle there, and the LP eaten by pigs, poultry and the farmer's family. In practice, this proved too complicated. Some of the fibre + 'whey' is fed to cattle immediately, the rest is ensiled, or dried, and sold. Bidkin is designed to be a research unit which will examine the economics of running a small-scale fodder fractionation unit: all products were to be consumed on site. Nevertheless, as a result of the changes which were found advisable, and with help from charitable organisations, most of the LP is now given to about 60 children, 5 to 8 years old. Some of the moist press-cake, mixed with twice its weight of wheat flour, is sold for 10 rupees kg^{-1} as infant food. Savangikar (1986) argues that, compared with equivalent food proteins, and making allowance for the β carotene and unsaturated lipids in it, LP (dry) is worth 26 rupees kg^{-1}. A unit of the type described by Butler & Pirie (p. 145), but made in India, was used to make the LP.

Again with help from charities and using a similar extraction unit, the same group has started a trial at Ellora. Three diets are being compared: a control diet in which protein is supplied by wheat, gram and groundnuts, and diets in which 12.5 and 47.5% of that protein is replaced by LP. Unpublished results show that growth is marginally better on the diets containing LP, and that these diets give significant increases in hemoglobin. Many different forms of presentation are being tried on tasting panels and on visitors. Coriander chutney, which is a popular relish in that part in India, is particularly interesting. Even LP from lucerne can be added at 15 to 20%; at 9% its presence is not noticed. Ellora is an excellent site for popularising LP because many visitors go there to see an elaborately carved set of caves.

In 1958, Kwame Nkrumah, the first Prime Minister of Ghana, became interested in LP, invited me to look at the potentialities and, as a result, invited Byers to spend 3 months studying the extractability of LP from plants available near Kumasi (Byers, 1961). Political changes then made the idea of LP production unpopular. Recently, with support from 'Find Your Feet' and other charities, and encouragement from the Head of State, Flt Lt

Rawlings, an extraction unit was installed by Dr P. Fellows at Kpone Bawaleshie, 34 km north of Accra. This resembles the unit described by Butler & Pirie (1981) except that the two sections are parallel rather than in tandem (a photograph is published, see Find Your Feet, 1985). A village cooperative manages the unit; it is still working out the economics of paying for cultivated plants and for a few species of wild plant collected locally, and charging for moist LP press-cake and fibre. Nearby villages are beginning to cooperate with Bawaleshie. This project, which depends entirely on local enterprise after getting the initial gift of an extraction unit, gets closer than any other project to the original idea of how LP will be most useful. Some interesting, and not wholly unexpected, pieces of local initiative have been added to the project. The 'whey', together with molasses, is fermented and distilled. It is maintained that the 'gin', mixed with LP, 'gives strength'.

Interest in LP started in Sri Lanka in 1966 and led to the purchase of an IBP pulper and belt press in 1970 by the Institute of Scientific and Industrial Research. Little use was made of it, and interest waned until, in 1981, 'Find Your Feet' and Sarvodaya realised that LP was a natural extension of the traditional culinary use of leaf juice in making 'kola kanda' (p. 138) (Davys, 1981). Sarvodaya is a local, non-profit-making organisation concerned with development and nutrition projects in 2000 villages. Only palatable leaves can be used in making 'kola kanda' and they contribute only about 1 g of protein in an average helping. There is less restriction on the use of LP: hence the advantage of using the more elaborate process. A juice extractor, similar to the one at Bidkin, was installed at Lunawa, 20 km south of Colombo. It is powered by three or four men pedalling – as if on bicycles. Sitting down to a job is more socially acceptable than standing up on treadles: but it is less effective. With optimal stroke and rhythm, a man can generate about 100 W continuously on treadles. The unit has about half the throughput of its prototype. It is being improved and other units are being installed. UNICEF supports this project. Extraction equipment has been sent to Egypt and LP production in small-scale units, for use as human food, is starting in Kenya with support from UNIFEM, the Womens' Fund of UNDP.

Carol Martin, chairman of 'Find Your Feet' (13–15 Frognal, London NW3 66AP) is the driving force behind almost all these feeding trials. She vividly describes her experiences in collecting

money and in overcoming bureaucratic obstruction and indolence
in 'All Grass is Flesh'.

Nutritionally valuable components other than protein

After protein, β carotene is the most important component of LP.
Tocopherol (vitamin E) is present, as is vitamin K (phylloquinone);
there have been few measurements of these components. Adequately washed LP contains little vitamin C, the B vitamins, or
calcium. Starch is a variable component (p. 58). Lipids are more
constant (p. 55), but little is known about the extent to which their
energy is metabolically available. Those which were extracted by
acetone and ethanol were not a substitute for half the soya bean
meal in poultry diets (Murai et al., 1984). Iron is always present:
the amount is variable and there is no evidence about the extent
to which it is absorbed.

The simplest and most sensible way to make use of the protein
and β carotene in leaves is to eat them in the normal manner as
a green vegetable. They cannot however be relied on to meet our
whole requirement. Although the association of greenness with
vitamin A activity has long been known (Hume, 1921; Dye et al.,
1927), it has only recently been stressed and epitomised in praise
for DGLV, i.e. dark green leafy vegetables. An extreme example is
the presence of 200 times as much β carotene in the outer leaves
of cabbage, which are usually discarded, as in the pale inner leaves
(Rothschild et al., 1977). Some leaves, e.g. nasturtium, contain 1
to 2 mg of β carotene g^{-1} DM: as much as is usually found in LP.
Daily eating of 30 g (wet weight) of the DGLV commonly used in
south India increased the serum retinol of children (Pereira &
Begum, 1968), and was as effective (Venkataswamy et al., 1976)
as massive doses of retinol in clearing the signs of xerophthalamia up
to the level of severity called corneal xerosis. Culinary skill is needed
to ensure good absorption of β carotene. By measuring fecal carotene
and assuming that there is no destruction in the gut (a reasonable
assumption that has been made by many since the studies of Hume
& Krebs (1949)), Lala & Reddy (1970) deduced that children
absorbed 57 to 93% of the 1.2 mg of β carotene in 40 g of
Amaranthus tricolor cooked in a little oil. Absorption by adults varied
(Rao & Rao, 1970); nevertheless, absorption from DGLV resembled
absorption from carrots or papaya. It seems likely therefore that the
carotene in LP, which is already finely dispersed, mixed with leaf

lipids, and liberated from the cell structure of the leaf, could have an important role in human nutrition.

Serum retinol was measured on some of the children taking part in the trial set out in Table 9 (p. 119). The values were 0.10, 0.23, 0.34, 0.39, 0.38 and 0.41 mg l^{-1} respectively in those being given no supplement, tapioca, horse gram, cereal + pulse mixture, skim milk and LP. It is generally agreed that 0.20 mg l^{-1} is the value at which signs of deficiency may appear. Clearly, all the supplements rich in protein increased serum retinol – presumably because they enabled retinol-binding protein, without which retinol cannot be transported in the blood, to be synthesised. I have already suggested (p. 117) that the better growth of children in Pakistan who were given LP, compared to those given milk, could have been a result of the β carotene in LP supplementing a vitamin-A-deficient home diet. These serum retinol measurements suggest that the home diets near Coimbatore (in South India) are not similarly deficient.

If all DGLV were of top quality, if they were not so widely disparaged as they now are, and if it were easier to persuade very young children to eat the rather bulky 20 to 30 g (wet weight) of DGLV that is needed to meet their vitamin A requirement, there would be no need to stress the value of LP as a source of β carotene. All reasonable steps should be taken to promote DGLV, but it is unlikely that they alone will, in the near future, meet the need. For example, Sommer *et al.* (1981), from a survey in Bangladesh, India, Indonesia and the Philippines, estimate that there are 0.5 M new cases of vitamin A deficiency annually in these countries: half of them become blind. Tarwotjo *et al.* (1982) and A. Pirie (1983) discuss the correlation between vitamin A deficiency and a smaller than average consumption of DGLV. Prevention of eye damage is not the only role of β carotene. Peto *et al.* (1981) and others (e.g. Anonymous, 1984) discuss sympathetically the part it may play in preventing cancer. When the place of LP in a food program is being discussed (Pirie, 1983) the significance of the β carotene in it should not be overlooked.

8

The value of the extracted fibre and the 'whey'

When a raw material is fractionated, there is sometimes little uncertainty about which fraction is the primary product and which the by-product. Sometimes, the by-product is treated as a waste because of technical difficulties, habit or laziness. More often, all the products of a fractionation are recognised as having value, but the value put on each may be arbitrary, or may depend on demand for that particular type of product at the time. Though not universally understood, these points have often been made before. Thus Lawes (1864) said: 'Linseed and other cakes are themselves, in one sense, manufactured foods. But the object of the manufacturer is not the production of cake, but of oil. If the farmer did not use the cake at all, it would still be made, and the oil would be sold for a higher price. As it is, the manufacturer makes the cake as a by-product, and the price he gets for it enables him to sell his oil so much the cheaper. But if manufactories were set up for the special purpose of preparing foods for stock, the whole cost of the undertakings must be charged upon the food...' This principle applies with particular force to fodder fractionation. The DM of an average crop is distributed between the three products – LP, fibre and 'whey' – in approximately the ratios 1 to 5 to 1. Different crops and methods of fractionation give different ratios, but it is obvious that uses must be found for all three products, and it is possible that each will be more valuable when separated from the others than in the original mixture. The value of LP as a food for nonruminants is the main theme of the preceding chapters. The value of the fibre may be enhanced, although it contains less protein than the original crop, because it is drier and has an improved texture. Furthermore, the process of LP extraction removes toxic and/or unpalatable components from material such as potato haulm and thus makes that

125

residual fibre a new source of ruminant feed. The components of 'whey' may be more valuable as substrates for microbial growth than they are as animal feeding stuffs in the original fodder. All the products of fodder fractionation must be used, so as to avoid local pollution as well as for economic reasons.

The patents taken out by Ereky (1927) and Goodall (1936) refer vaguely to the conservation of the fibre from which they had extracted juice. Better silage can be made from lush crops if they are wilted and bruised so that there is less loss from seepage (e.g. Derbyshire *et al.*, 1969), and quicker compaction of the mass: it is obviously more economical to conserve a crop by drying if pre-treatment has removed some of the water. During the 1930s several attempts were made to 'dewater' crops with various commercially available machines – including the now-fashionable twin-screw expeller. They were all unsuccessful because of inadequate appre-ciation, shown by repeated use of such phrases as 'pressing out juice', of the fact that pure pressure gets little juice from an unrubbed leaf. Because the machines used did not apply pure pressure, there was some rubbing; juice was therefore liberated, and protein was liberated along with it. This protein was recovered by Dawson (p. 14) and Goodall (1950) but was wasted by Randolph *et al.* (1958) and Casselman *et al.* (1965). If oil prices had increased at that time, and if there had then been the present sensible concern with waste of energy, the primary product of fodder fractionation would have been the fibre: LP would now be considered the by-product. In Europe and the USA this may already be the situation. Failure to appreciate the value of the fibre explains various gloomy guesses at the cost of LP production which were made 15 to 25 years ago.

Fodder fractionation enhances the value of an existing crop, it also gives farmers an incentive to increase productivity. When forage will be used simply as feed for ruminants, there is little incentive to fertilise, harvest and irrigate it so as to get maximum yield of DM. Such a crop would be too lush to be made into hay, it may be too strongly buffered to make silage (Wallace, 1975), it would be extravagant to dry it, it would contain more protein than a ruminant needs and the excess protein may cause bloat. The proposition that a policy of LP production could increase the supply of cattle fodder, because it would give a motive for maximum DM production, was thought paradoxical when first stated (Pirie, 1942*a*). It now gains support (e.g. Wieringa, 1983).

In Britain, 39% of the permanent grass gets no fertiliser (National Economic Development Office, 1974); most of this land is probably too steep or rough to be mown, but much of the grass on the land that is, or could be, ploughed would yield more DM if more adequately fertilised. A farmer can make full use of a crop that contains more protein than a ruminant needs, by mixing it with a digestible but protein-deficient material such as beet pulp or alkali-treated straw. But the lushness of a well-fertilised crop would remain a problem if fodder conserved by drying were required. This problem becomes particularly acute when a drier is operated continuously so as to make maximum use of equipment. A crop may then have to be processed although it has dew, or even rain, on it. In these conditions, mechanical removal of water is essential for economy.

Most of the crops that are dried commercially contain 25 to 14% DM; with surface water they may contain only 7%. When the DM is as small as 10%, direct drying becomes uneconomic and the crop is allowed to wilt in the field before being dried. There may be little loss if, by chemical treatment, the standing crop is wilted before being mown. If it is mown, left on the ground to wilt, and then collected, there is considerable loss because of incomplete collection (e.g. Gordon *et al.*, 1969). To put these figures in perspective: to get 1 t of DM from crops initially containing 7, 10, 14 and 25% DM it is necessary to remove 13, 9, 6 and 3 t of water, respectively. Fibre coming from an efficient LP production unit contains 30 to 40% DM. To get 1 t of DM from such materials, only 2.3 and 1.5 t of water must be removed. During periods of fine weather it is possible to remove that amount of water by drying in unheated air. The final product of any form of commercial drying contains 10 to 14% of water; if calculations are made on that basis, the apparent advantage of putting pressed fibre into the drier becomes even greater.

Early interest in fodder fractionation was aroused by the possibility that half to three-quarters of the evaporation load could be removed from the drier, and this possibility was part of the justification for repeated reemphasis of the potentialities of such processes (e.g. Pirie, 1953, 1966*b*). It may not be possible, with existing equipment, to produce fibre that consistently contains as much as 40% DM, and it may not be advisable to do so because of the approximate proportionality between the removal of water and the extraction of protein. LP is more valuable as a human food, or feed for nonruminant animals, than protein left in the fibre. There are here three conflicting factors: a small extra yield of LP costs more for energy and wear and tear of equipment than the main yield; if the fibre is to be dried it is advantageous to get as much juice out as possible; so long as the nonprotein part of the fibre is reasonably digestible in ruminants, fibre with depleted protein can be supplemented with other sources of N such as urea.

It may be worthwhile reemphasising some points which have already been made (pp. 19, 54). Juice running out from pulp under gentle pressure has a greater concentration of protein in it than juice coming out when pressure is increased. There is an inverse relationship between the time taken for pressed fibre to reach a given percentage DM and the thickness of the fibre. It may therefore be best, in commercial production, to press in two stages: first gently in a thin layer to get juice from which to coagulate LP, and then more intensely to get fibre which can be economically dried. The amount of LP that can be recovered from that second press-juice may not justify the cost of coagulating it. But pressing out water costs less than evaporating it.

The quality and use of the fibre

All the properties of the fibre depend on the quality of the crop that was processed and the thoroughness with which it was extracted. Byers & Sturrock (1965) found 0.74 to 3.3% N in the fibres from 17 crops harvested at different ages and fertilised at different levels. Fibre from hybrid napier grass (gajraj) contained 1.6 to 2.2% N (Gore *et al.*, 1974) and from berseem 1.9 to 2.8% (Mungikar *et al.*, 1976a). The larger values were given by crops that initially contained the most N, but there was no strict proportionality. During the extraction of LP, salts, carbohydrates and other substances are extracted. Consequently, if half the protein is extracted from a crop, the N content of the fibre is more than half that of the original crop. Table 10 shows the extent to which N was removed from several crops by extraction of LP.

For three reasons the value of the fibre as a feed for ruminants is likely to be greater than the value of a crop that initially contained the same amount of N as these samples of fibre. The fibre is made from younger and less lignified leaves, a larger fraction of the N in the fibre is true protein, and comminution of the fibre increases its palatability. Furthermore, ruminants make less efficient use of soluble proteins, which are hydrolysed and partly deaminated in the rumen, than of protein that, for various reasons, is insoluble and by-passes the rumen (Chalmers & Synge, 1954; Kempton *et al.*, 1977). This gives fibre-bound protein an added merit. In the 1950s we found that cattle ate the fresh fibre readily. This point has been stressed by Greenhalgh & Reid (1975), Houseman & Connell (1976), Connell & Houseman (1977) and Wieringa (1983). Sheep eat it readily and ruminate normally (McLeay *et al.*, 1982).

The many experiments in which the performance of animals fed on fibre has been compared with that of animals fed on the original crop have given varied results. Among the factors which may explain this variability are: differences in the extent of LP extraction, differences in the extent to which other nutritionally valuable leaf

Table 10. *Nitrogen contents of the original crop and of extracted fibre made from it*

Crop	Original percentage of nitrogen in the dry matter	Percentage of nitrogen in the fibre	Reference
Lucerne	3.1 to 3.5	2.5 ⎫	Hartman *et al.*
Pea vine	2.0	1.55 ⎭	(1967)
Average of five grasses	2.9	2.2	Maguire & Brookes (1973)
Lucerne	3.5	2.7	de Fremery *et al.* (1974)
Lucerne	4.8 to 2.5	4.3 to 2.4	Connell (1975)
Ryegrass (average of five harvests, May to October)	2.6	2.0 ⎫	Connell & Houseman
Lucerne (early)	3.3	2.4 ⎬	(1977)
Lucerne (late)	3.2	2.5 ⎭	
Pea vines	2.75	2.36	Gonzalez & Alzueta (1984)
Maize	1.8	1.7 ⎫	
Sorghum	1.8	1.7 ⎪	
Hybrid Napier	1.3	1.1 ⎬	Kasture *et al.* (1984)
Cowpea	3.5	2.2 ⎪	
Dolicos	3.2	2.2 ⎭	

components, e.g. β carotene, calcium and phosphorus, were extracted, and damage by overheating if the fibre had been dried. The last factor may explain the 10% loss in digestibility usually found after drying (Vartha *et al.*, 1973; de Fremery *et al.*, 1974; Greenhalgh & Reid, 1975; Houseman & Connell, 1976). At one extreme come experiments in which bulls gained more weight when fed *ad lib.* on fibre than on grass (Eidrigevich *et al.*, 1978; Strazetelski *et al.*, 1981), cows in one year out of four gave more milk on fibre than on grass (Connell & Foxell, 1976), and cattle grew faster, with greater conversion efficiency, on fibre than on the original crops (Jones, 1983). Some trials (e.g. Wieringa, 1983; Fujihara & Ohshima, 1984) demonstrated that the fibre was satisfactory fodder for cattle or sheep although strict comparisons with the original crops were not made. Ahmad *et al.* (1980) found that calves ate the residue from berseem less readily than cotton cake but, for equal weight gain, it cost less. At the other extreme are experiments (e.g. Donnelly *et al.*, 1980; Stockdale *et al.*, 1981; Bryant *et al.*, 1983) in which milk yield was 5 to 10% less on fibre. This diminution in milk yield seems to be a feature of fibre from grass rather than lucerne (Stockdale *et al.*, 1981). As has already been pointed out (p. ix), LP can be made from plants which are toxic or for other reasons are inedible. The processes of pulping and pressing should make it possible to prepare fodder from such plants as well. This line of work has not been actively pursued, but diets containing 50% of fibre from water hyacinth are satisfactory feed for sheep (Borhami & El-Shazly, 1984).

Fibre has been dried satisfactorily in various types of commercial grass drier. Dry material is a convenient component of cattle feed, but even after part of the juice has been removed, drying is still expensive and is often an extravagance. As soon as there was regular LP production at Rothamsted we made silage, with fibre from several species, in concrete drain pipes (60 cm diameter) and in smaller vacuum packs. Enough 'whey' was added to fibre to saturate it, but not enough to produce any effluent. The precise quantity added did not seem to be critical and cattle ate the silage readily. Silage was made from residual fibre from beet, carrot (*Daucus carota*), beans, potato and lucerne (Oelshlegel *et al.*, 1969), berseem (Magoon, 1972), hybrid Napier grass (Mungikar & Joshi, 1976) and pea vines (Nørgaard Pedersen *et al.*, 1980). Ohshima & Kogure (1984) made good-quality silage from nine species provided the fibre was dry enough and contained enough fermentable carbohydrate. The excellent quality of silage from fibre left after LP extraction was commented on by Nørgaard Pedersen (1983) and Lu *et al.* (1980); the latter group found that lucerne silage had the normal retention time in goats. Substitution of silages from sorghum, ryegrass or oat fibres for conventional silage from these crops did not affect milk yield (Ohshima & Sogo, 1984).

Nevertheless, there were some failures, e.g. Raymond & Harris (1957) and Vartha *et al.* (1973) until they added molasses. Presumably success in the early trials at Rothamsted was the result of the addition of fermentable carbohydrate in 'whey' as well as of the exclusion of air. The technique is now widely adopted, e.g. Dakore & Mungikar (1985). Sugar cane tops, which Kasture *et al.* (1984) find a beneficial addition to fibre silage, may act by supplying fermentable carbohydrate. Removal of some of the protein and cations diminishes the buffering capacity of the original crop (Wallace, 1975; Ohshima & Oouchi, 1979) and thus enables a smaller amount of carbohydrate to effect the diminution of pH needed for good silage. Quality can be improved by adding formic acid or formaldehyde (Fujihara & Ohshima, 1980). Juice separation can remove unfamiliar or unappealing flavours. Thus goats preferred potato haulm silage to lucerne silage (Ream *et al.*, 1983). But this is not a panacea: silage from water-weed fibre has been rejected.

Cattle are accustomed to diets containing much carotene, and there is evidence that carotene deficiency leads to infertility. (Cooke, 1978). Although most of the carotene associated with LP is stable when the LP is properly processed (p. 80), the carotene in the fibre is rapidly destroyed (Arkcoll & Holden, 1973). Whether the fibre is being preserved by drying at a high temperature or by making silage, if the product will be an important component of the total fodder given to cattle, it may be advisable to start conservation quickly so that enzymes are inactivated by heat or prevented from acting by the absence of oxygen in packed silage.

When broad-leaved forages are dried and then milled gently, the protein-rich leaf blades break up more completely than the more fibrous parts which contain little protein. By sieving, fractions can therefore be made which contain twice as much protein as the original dried forage (references in Pirie, 1977). This process could be used to make feed for nonruminants from the fibre residue after

LP extraction. The protein-depleted fraction would then be of little use except for making paper or cardboard, as a fuel, or as a substrate for methane fermentation. The last two uses are sometimes suggested for the unfractionated fibre: these would be wasteful ways of using a potentially valuable fodder.

The quality and use of the 'whey'

The deproteinised juice from an LP production factory, which by analogy with cheese manufacture is often loosely called 'whey', would have the Biological Oxygen Demand of the sewage from a small town. It would therefore be extravagant to discharge it into a sewer or treat it so that it can be discharged into a stream. Of the three products of fodder fractionation, this is the most variable; the DMs of 61 samples from nine species ranged from 11 to 47 g l^{-1} (Festenstein, 1972): many other publications (e.g. Hill-Cottingham & Lloyd-Jones, 1979) give similar results. If crops harvested in wet weather had been included, the range of DMs would have been still greater. When LP is being made as a human food, the water used to wash it would be added to the 'whey' and would nearly double its volume. A comparable amount of water would be used for cleaning the equipment, and that might be added also. The values suggested in the following paragraphs are based on analyses of true 'whey' and do not make allowance for this probable dilution.

The concentrations of the various substances dissolved in 'whey' span an even larger range than the DMs. Thus the N contents were 0.25 to 1.2 g l^{-1}, and the total carbohydrate contents were 2.5 to 22 g l^{-1}. The true protein content of correctly heated and filtered 'whey' is negligible. Festenstein (1972) measured the contribution made by glucose, fructose, xylose, sucrose, fructose oligosaccharides and fructosan to the total carbohydrate; there were great seasonal and species differences. Thus glucose was a negligible component of the total in 'whey' from ryegrass harvested in August, but made up 28% of it with wheat in June; at that time, 13% of the carbohydrate in 'whey' from wheat was sucrose, but it was never found in mustard. Allowance for this variability must be made when suggesting uses for the 'whey'.

The simplest way to use 'whey' is to carry it back, as a return load, to the fields from which the crop was harvested. About half the wet weight of the crop is in the 'whey', it will therefore contain nearly half the potassium and a smaller fraction of the phosphorus in the crop. 'Whey' from a leguminous crop may contain a quarter of the crop N; the fraction is smaller with cereals unless they are harvested soon after being fertilised. No systematic analyses for these elements have been published; their actual value as fertilisers can only be guessed. If 'whey' were restored evenly to the full area from which the crop came, it would form a layer only 1 to 3 mm deep and would supply a substantial part of the potassium needed for the growth of a second harvest.

As is well known, germination and plant growth may be inhibited on land heavily contaminated with silage effluent. Some plant extracts have a general phytotoxic action; with others, phytotoxicity is more specific (e.g. Marchaim *et al.*, 1972; Allison, 1973; Moore, 1976). There was therefore some fear that 'whey' would be harmful. If returned to an area much smaller than the area from which the crop came (the conditions in the neighbourhood of a silo), it inhibits germination and may 'scorch' young leaves. So far as we know (Arkcoll, 1973*b*) it is harmless when returned to as little as a fifth to a tenth of the area from which it came. This was also the experience of Ream *et al.* (1977). By encouraging the growth of gas-forming bacteria such as *Clostridium butyricum* and *Clostridium pasteurianum*, the carbohydrates in 'whey' can improve the structure of intractable soils; this effect persists after the gases escape because of soil stabilisation by bacterial polysaccharides (Arkcoll, 1973*b*).

Most of the components of 'whey' are of doubtful nutritional value for nonruminants, and 'whey' is so dilute that it would be difficult to dispose of all of it in ruminant feed in its original state. Hollo & Koch (1971) concentrated it *in vacuo* and mixed the resulting paste with the LP. The advantage of complicating the process in this way, rather than spray-drying the whole juice as Hartman *et al.* (1967) suggested, is that the 'whey' does not coagulate on heating and so can be concentrated in a thermally efficient, multiple-effect or vapour-compression unit. In spite of the economy thereby attainable, it seems unlikely that it would be worthwhile evaporating 'wheys' at the more dilute end of the concentration range. If the more concentrated 'wheys' are being evaporated, the product should not be added to the LP but to the fibre so that it is used as ruminant feed. This procedure is now used by Koch (1983) and, as papers in the Aurangabad Symposium (N. Singh, 1984) show, it is also used in France, Italy, Japan and

New Zealand. When 10% of a cattle diet consisted of concentrated 'whey' rather than molasses, growth rate increased (Bris *et al.*, 1970).

The rapidity with which 'whey' stinks when left for one or two days in summer shows that it is a good microbial culture medium; if used in that way it would not need to be concentrated. This use was suggested as soon as systematic work started (e.g. Pirie, 1951) and samples were sent to various organisations interested in the cultivation of microorganisms, but its potentialities have not yet been thoroughly investigated. That cannot be expected until a regular supply is assured. Jönsson (1962) compared as substrates for seven microorganisms, 'whey' from pea vines with four other agricultural by-products. *Rhizobium meliloti* grew particularly well on it; when an extra source of N was added, penicillin production was satisfactory. Rhizobia grow well on lucerne 'whey' (Gangawane & Nehemiah, 1980).

Later studies have been concerned with the production of 'biomass' that could be used as human or animal food. The 'whey' from 1 kg of water hyacinth yielded 12 g (DM) of a strain of *Saccharomyces cerevisiae* isolated from rotting hyacinth (Oyakawa *et al.*, 1968). This may show the advantage of using a yeast strain already adapted to the substrate. Butt *et al.* (1972) got only about half that yield from nine strains grown on 'whey' from berseem, mung bean (*Phaseolus mungo*) or guar (*Cyamopsis psoralioides*). Paredes-Lopez & Camagro (1973) found four yeasts that grew well on lucerne 'whey' and three that did not. The sample of lucerne 'whey' used in experiments in New Zealand (Barnes, 1976) still contained 3 to 5 g of protein l^{-1}. Both *Saccharomyces* sp. and *Rhodotorula* sp. converted half of that, and of the 0.9 to 1.5 g of free amino acids l^{-1}, into 5 to 6 g of 'biomass'. The carbohydrate in the 'whey' was used up completely. Results with *Aspergillus, Pseudomonas, Flavobacterium* and *Corynebacterium* were similar. Worgan & Wilkins (1977) quote an experiment showing that *Fusarium semitectum, Trichoderma viride* and *Aspergillus oryzae* removed 65 to 78% of the Chemical Oxygen Demand from lucerne 'whey'. A succession of organisms will probably make even fuller use of the components of 'whey'. Thus Kümmerlin (unpublished) got 12 g of 'biomass' from *Candida utilis* followed by *Endomycopsis fibuliger* growing on a litre of 'whey' from tall fescue (*Festuca arundinacea*); the yield was smaller when the sequence was reversed. In Hungary, a litre of 'whey' (species unspecified) yielded 25 to 30 g (DM) of yeast (Koch, 1974). The yield of dry yeast from lucerne 'whey', made during large-scale production, was 18 to 22 g l^{-1} (Koch, 1983). As would be expected, yeast grown on 'whey' is as good as feed for weanling pigs as fish-meal (Ohshima & Ueda, 1983).

'Whey' from many different species has been tested. Kümmerlin (1984) got good yields of *Candida utilis* on 'whey' from ten species, including water weeds and maize taken at the time of harvest of the ears. Ethanol production with *Saccharomyces cerevisiae* was less successful. 'Whey' from turnip, mustard and radish yielded ten times as much yeast as from cauliflower, spinach and winged bean (*Psophocarpus tetragonolobus*) (Chanda *et al.*, 1980). No great reliance should be placed on these results because of the variability of the composition of 'whey'. In several experiments, scarcity of some components of 'whey' have been corrected. Thus Chanda (1983) improved penicillin production by adding sucrose: Mudgett *et al.* (1980) improved the yield of *Candida* by adding N and phosphorus.

When supplies of 'whey' become regular, and as experience increases, variability will become less of a problem because differences between harvests will become predictable from an extension of the work of Festenstein (1972) and Nowakowski & Byers (1972) who studied the effect of different fertiliser regimes on the non-protein N of ryegrass. The necessary additions, usually N and phosphorus, needed to produce a balanced medium can then be made by rule of thumb. Even if the primary object of microbial cultivation is to diminish the Biological Oxygen Demand, some chemical additions will probably be needed because 'whey' usually contains excess carbohydrate. In spite of these difficulties, the active interest taken in 'whey' as a microbial substrate is shown by brief comments in publications from Brazil, Byelorussia, Egypt, Japan, Latvia and Poland.

These results serve only to illustrate a potentiality. It would be reasonable to use an organism, such as *Candida*, that can hydrolyse polysaccharides. For practical convenience, funguses that form mycelia have advantages because the 'biomass' can be collected by filtering rather than by centrifuging. *Paecilomyces* and *Rhizopus* seem not to have been tried. The latter is particularly worthy of trial because it is already eaten extensively in South East Asia. Furthermore, fungus mycelia tend to contain less nucleic acid than yeasts.

Plant extracts comparable to 'whey' are well known as sources of substances with proved pharmacological activity. Nevertheless, claims that 'whey' stimulates plant growth to an extent that cannot easily be explained as the result of the presence of N, phosphorus, potassium, etc., or that 'whey' contains 'unidentified growth factors' for animals, may be treated with some scepticism. The roster of physiologically active molecules is probably far from complete, but mortality among claimants for inclusion in that roster is large. For example: 11 authentic B vitamins no longer precede vitamin B12. It is, however, worth recording that the triacontanol in lucerne extracts stimulates the growth of several plants (Ries *et al.*, 1977). McClure (1970) gives an enormous list of substances, many of them pharmacologically active, in water weeds; and water hyacinth 'whey' may stimulate plant growth.

When the 'whey' from a certain species contains a substance of proved efficacy, as a growth stimulant, it could be used as a spray. The exploitation of 'whey' as a source of purified substances depends on those substances having properties that are sufficiently unlike those of unwanted constituents of the juice for separation to be possible by commercially viable techniques. Attempts to separate ascorbic acid from lucerne juice failed several years ago because the ratio of ascorbic acid to the total quantity of organic acid collected on ion-exchange resins was unfavourable. However, if potato haulm were being used as a source of LP, it would be simple to precipitate glycoalkaloids (e.g. solanin and related substances), and little else, from the 'whey' by adding alkali. There is at present little demand for these substances – that was at one time the situation with the steroids in agaves and yams which are now important as starting materials for pharmacological syntheses. Alkaloids can be separated either by precipitation or by extraction into a solvent immiscible with water. Skinner (1955), Mitscher (1975) and Roja & Smith (1977) list many plants with antibiotic activity – though roots rather than leaves were usually studied. It is even possible that dyes, such as indigo and madder, could be by-products of LP production and could emulate rubber by competing with synthetic material. Therefore, when plant species or varieties are being screened as possible sources of drugs or dyes, it would be worthwhile paying some attention to the potentialities of their leaves as sources of LP.

These suggestions may seem like 'turning the clock back'. It is worth bearing in mind that industrial chemistry prevailed over extraction from 'natural' sources at a time when a science-based industry contended with traditional, or even peasant, agriculture. The balance may sometimes tilt the other way when agriculture also has a scientific basis. Furthermore, there are now proposals that photosynthetically generated 'biomass' should be used as a source of energy for heavy industry. If that proposal were followed up, some of the energy would be used to make the intermediates needed for synthesising the required compounds. It might be simpler, and would certainly be more logical, to rely on plants to do more of the synthesising.

9

The role of fodder fractionation in practice

Evidence set out in the preceding chapters shows that production of LP in bulk is technically feasible and that the product is nutritionally valuable. Stress was laid on the importance of considering the whole fractionation process without assigning pre-eminence to any one fraction. It is worth recording that in about 1960 a research group in Florida was collecting the fibre and discarding the juice; a group in Uppsala was collecting the 'whey' and discarding the fibre and LP; in Rothamsted we collected the LP and discarded the fibre and 'whey' but emphasised in all publications that this was a temporary state of affairs. There is nothing novel in the idea of fractionating plant material. Cereals are separated from chaff and often from bran. Potatoes are peeled. Sugar and oil are extracted from various crops. The petioles and outside leaves of many vegetables are rejected. It is odd, in the light of the last example of separation or selection, that it has taken so long for the idea of fractionating forages to gain acceptance. We have relied for too long on techniques which are within the capacity of moderately intelligent monkeys and squirrels.

Unfractionated leaf juice is used in a few communities – notably in Sri Lanka (Davys, 1981). There, kola-kanda is made by pounding leaves and coconut in a mortar, squeezing out the juice, reextracting the residue with water, and adding the combined extracts to boiled rice. About 30 g of leaf is used for each helping, leaf will therefore contribute about 1 g of protein. Because there is no separation of LP, only palatable leaves can be used. So far as I know, no community takes the further step of coagulating leaf juice and using the curd: I would welcome evidence that that is a traditional technique somewhere. The technique has been available for millenia – it is no more difficult than preparing sago or washing the

138

cyanide out of tapioca – but seems not to have been adopted. Inedible leaves were therefore subjected to conversion in animals rather than to fractionation.

The reasons for thinking that more attention should be paid to crops yielding leaves rather than seeds or tubers, and that leaf crops should be fractionated, were stated in the Introduction (p. viii) and need not be repeated. There are many of them, and many situations in which fodder fractionation would be advantageous. Scepticism is aroused when so many claims are made. An advocate is often asked whether factory, farm, or family operation is envisaged. The reply that all three procedures are suitable and that choice depends entirely on the particular situation seems to some sceptics as unconvincing as the advertisement for a panacea. Similarly, there is often argument about the relative importance of LP as a human or an animal food, and about the extent to which it is advantageous to deplete the fibrous residue of protein. Again, the answer depends on circumstances. In a poor country with poor communications, where it rains nearly every day, and where protein sources are scarce, it would be advisable to extract as much LP as possible and eat it near where it is being made. In an affluent country with a long winter during which cattle need fodder, attention would be paid to the partial dewatering of the fibre rather than to the extraction of LP, and most of what was extracted would be fed to pigs or poultry.

There will be little dispute about the value of a policy of fodder fractionation when the starting material is at present abundant and unused, e.g. a water weed or a by-product. There may be doubts about the feasibility of collecting material from water, or from the scattered sites at which by-products accumulate, but there will be no doubts about the advantage of making fodder and LP from an unexploited resource if that is feasible. But sites where a supply of weed can be regularly collected are, fortunately, rare, and by-products are intermittent. The main questions that arise concerning fractionation techniques are: 'Should a crop that is already being used unfractionated, be fractionated; and should a crop be grown for fractionation in place of a crop grown for conventional use?' These questions can be considered either in terms of the expected yield of edible proteins, or in terms of profitability.

Local production and use

The simplest situations to consider are communities in which most of the food is produced locally for local consumption. If, because of a local resource such as fish, protein is not scarce, there would be little advantage in making LP. There, if food is scarce, energy will be the main requirement, and cassava or yams are a more convenient source of energy than a leaf crop. In arid, or semi-arid regions the case for LP is uncertain. Crops grown with irrigation, but in very dry air, tend to have rather dry, leathery leaves from which protein does not extract as easily as it does from lush leaves, and it is obviously easy to conserve fodder by drying when the air is dry. It is in humid regions where protein deficiency is the main nutritional problem that the case for partially substituting LP crops for cereals, or even for protein-rich crops such as beans, is strongest. The evidence assembled in Chapter 3 shows that LP can give a greater yield of protein than any other crop. In regions where it rains nearly every day, so that it is difficult to ripen seed crops, the case is overwhelming. These are the regions where the first trials of family, or village, production of LP should be made. So many people live in such regions, that LP would make a contribution even if it were used nowhere else, and it is obviously wise to start at sites where the probability of success is greatest.

Choosing sites for getting work started on the use of LP as a normal food is as important as choosing climatically suitable regions for the work. Obviously, local enthusiasm is essential; without it, a project will probably collapse when those who come to advise, and to demonstrate equipment, leave the site. It is convenient for these advisers if the chosen sites are near a town and on good roads. But those most in need do not live in towns. There there is access to donated food and to other services. Work can usefully start at such sites: as soon as possible it should move out of town down 50 km of unmade road. There is now ample evidence (e.g. Mason *et al.*, 1985) that that is where need is greatest. What people living in such sites need is simple, robust equipment, preferably not dependent on electricity. Ideally, it should be possible to copy the equipment using locally available scrap. It should be able to produce from 2 to as little as 0.2 kg of LP daily. In spite of successful production and use of LP in Ghana, India, Mexico and Sri Lanka, the design of an ideal unit still remains a relatively neglected aspect of research on LP. It will take several decades of social and political progress to abolish the need for equipment of this type.

The potential advantages of finding out how to make LP, and how to use it in suitable regions, seem so obvious that outside observers are puzzled by the slow progress of work on the subject. I have discussed elsewhere the general problem of opposition to innovation – nutritional or otherwise (Pirie 1971*b*, 1972, 1975*b*). A specific obstacle is that thoughtful people in poor countries are understandably suspicious of the proposal that they should start using something that is not used elsewhere. The fact that climates and circumstances are different may not convince sceptics completely. Acceptance has not been helped by the tendency of various organisations, e.g. FAO (1964, 1969), United States President's Science Advisory Committee (1967), United Nations (1968), to accompany a phrase such as '...the potential value of such products is unquestionable...' with a catalogue of all the possible difficulties that may be encountered with LP. Difficulties undoubtedly exist: it is unfortunate that it is only for LP that they are itemised although they obtrude to a similar extent with all other novel protein sources. A foolish and tendentious statement by the Protein Advisory Group (1970) asserted that village production of LP was impractical. A more sensible statement from the same group was given only restricted circulation.

During the past few years there has been a welcome change in the 'climate of opinion'. There is now research on LP in about 50 countries. As the references in this book show, activity is now greatest in Japan; there is work there in ten institutes. At least four institutes in China undertake work on LP, but I have not seen any publications from there. Three comprehensive symposia volumes on LP have appeared (Telek & Graham, 1983; N. Singh, 1984; Tasaki, 1986) and the advantages of fractionating some crops have been given sympathetic consideration in symposia on the broader aspects of agriculture (e.g. Jollans, 1981; Griffiths & Maguire, 1983).

Commercial production

The gloomy conclusions come to by some of the organisations listed in the last paragraph but one did not deter some animal feeding-stuffs manufacturers from taking an interest in the potentialities of fodder fractionation. Part of the reason for the failure of early attempts at commercialisation (p. 14) was the use of unsuitable equipment. This was probably also part of the reason for the ending of production by Batley-Janss (Brawley, California) and by Technion (Haifa). A contributory factor in these failures was the absence, at that time, of emotional commitment to energy conservation. France-Luzerne (Gastineau, 1974, 1976) started research on fodder fractionation in 1956 and started large-scale work at Chalons-sur-Marne and Mairy-sur-Marne in the 1970s. Daily production of dry LP had reached 2 t by 1974; annual production is now 7000 t (Gastineau & de Mathan, 1984). France-Luzerne puts particular emphasis on the xanthophyll in LP, which is mainly used in feed for poultry, and on the fuel saved when fibre is dried. A factory is being built in California with an expected annual production of 20000 t of LP.

As a result of many years work in Wisconsin (Koegel & Bruhn, 1977; Ream *et al.*, 1983) leaves are pulped by forcing them through holes 16 mm in diameter, as in a 'California' or 'pelleting press'. Juice is then squeezed out in a 'double cone' press. Řezníček & Truxová (1984) extruded leaves through plates with holes from 2 to 20 mm in diameter and found greater liberation of cell contents with the smaller holes. The power needed for extrusion was not measured: it may increase disproportionately as the holes get smaller, so that there is an optimal size. The two studies differ: in Wisconsin there is shear as well as pressure, whereas Řezníček & Truxová applied pressure alone. My conclusion (p. 10) that pounding was ineffective, coupled with some other experiments, suggest that that difference is important, and that extrusion takes less power, and is more effective, when combined with shear. Bruhn and his colleagues, in the papers quoted above and elsewhere, make many sensible comments on the importance of simplicity and cheapness; they point out that the traditional approach of those who make industrial equipment causes them to charge ten times as much as is charged for agricultural equipment of the same weight. That was also my experience when several aircraft firms, wishing to diversify 20 years ago, came along with fantastic designs for LP equipment. Anyone about to start work on large-scale production would be well advised to read the papers by McDonald *et al.* (1986) and Shearer (1986) which describe problems arising in New Zealand from abrasion, foam and fouling.

LP has been produced for many years in Hungary (Hollo & Koch,

1971; Koch, 1983), and there are three production units in Japan (Ohshima & Ueda, 1984), a large unit in Denmark (Christensen, 1984) and production units in Italy, Spain, the USSR and possibly other countries. It is sometimes difficult to get precise information because of the combined effects of commercial optimism and secrecy.

Forage suitable for fractionation is most abundant in the northern hemisphere in May, June, September and October: cereal straw is available in August. Now that there is a revival of interest in increasing the digestibility of straw by treating it with alkali, and the clumsy techniques of soaking in dilute alkali are being replaced by treatment with small amounts of concentrated alkali, it is apparent that fodder fractionation and straw treatment dovetail neatly together. Many of the facilities of a factory could be used for both processes. One attempt to alternate the two processes (in Yorkshire, UK) has been abandoned: perhaps because of the choice of unsuitable processing equipment. But the principle is sound and it should be exploited again with careful planning and attention to the efficiency of the equipment used to pulp and press the crop.

The advisability of starting fodder fractionation in a region depends on the climate. If fine weather can be relied on in summer, so that partial field drying is reliable, and if the LP would be used only as broiler feed, there seems to be little advantage in extracting LP rather than separating the partly dried crop into protein-rich and protein-poor fractions mechanically (cf. Pirie, 1977). The strength of the case for large-scale fodder fractionation by methods of the type described in this book depends, in any region, on the unpredictability of summer weather, and on the demand for conserved winter fodder.

Technical factors which affect economic assessment

There have been at least 20 attempts to assess the economics of fodder fractionation. In the first edition of this book, they were given fairly close analysis. Recently the general trend of economic assessment has been more sympathetic but, for many reasons, it does not seem to be worthwhile repeating that exercise. The overriding reason is that the assessments reflect the initial prejudices of the assessor, or guesser, rather than the realities of the situation. There are so many factors in each situation that there are ample opportunities for prejudiced weighting. I hasten to add that prejudice is nearly as often shown in the rosy as in the gloomy assessments. In spite of my sceptical opinion that, insofar as economics is a science at all, it is an experimental science, some of the more cogent assessments may be listed (Dumont & Boyce, 1976; Morris, 1977; Wilkins *et al.*, 1977; McDougall, 1980; Bray, 1983; Enochian *et al.*, 1983; Schoney & McGuckin, 1983; Ikumo & Takigawa, 1984; Savangikar, 1986). Economic argument cannot be rigid because it is affected by many factors, such as those listed below.

The crop. The costs of growing a crop are not affected by the use made of the crop except for the increased service that may be got from equipment because fodder fractionation depends less on weather than conventional conservation.

It may be worthwhile using fertiliser and irrigation water at levels which increase the yield of both protein and DM because, with fodder fractionation, the resulting fodder does not contain more protein than a ruminant can use effectively. There is thus a better return on the capital value of land.

In principle, a waste or by-product costs nothing – one could even be paid for removing water weeds or vegetable residues. The cost of transport would, however, usually be greater than the costs of on-the-farm transport, and the amount of protein extractable would invariably be less than the amount extractable from an equal weight of a crop. Processing equipment would therefore be used less efficiently on wastes.

Harvesting. Flail-type harvesters are simple and economical to run, but they suck up an amount of dirt that would be unacceptable if human food were being made, and the crop is so battered that it cannot be washed before pulping. Furthermore, presumably because of autolysis in the battered crop, LP yields are smaller from a flailed crop than from a crop cut with an old-fashioned reaper (Rothamsted, unpublished). There is more protein in the leafy upper part of a crop than in the stem lower down and, with many crops, regrowth is faster if the crop is not cut near the ground. Extraction equipment would therefore produce more LP in a given time if it were fed with the upper part of a crop only. It is difficult to assess the conflicting economic effects of increasing the cost and difficulty of harvesting, so as to make later stages in the process more efficient.

Pulping. High-speed pulpers, i.e. those with beater tip speeds of 40 to 60 m sec^{-1} can cope with crops of more varied texture, and get out more protein from them, than other types of pulper. They were, and still are, the best units for agronomic work. There is general agreement that they are obsolete for bulk production. For large-scale production the extrusion arrangement used in Wisconsin (p. 142) is probably the best: it is not easy to see how that arrangement could be scaled down for domestic or village use.

An effective unit (Butler & Pirie, 1981) for small-scale production retains the simplicity of the screw expeller, but the section in which pressure is applied is preceded by a set of angled paddles which rub the crop thoroughly before pushing it into the section in which pressure is applied. A unit which processes 1 to 2 kg min^{-1} (wet weight of crop) takes 300 W. Several of these extractors, with minor modifications, have been made in India. A more radical modification, with the rubbing and pressing sections in parallel rather than in tandem, is made in Sweden (p. 122). A smaller version (Pirie unpublished), hand-driven with pedal assistance, is suitable for domestic use.

High-speed machines waste energy by creating wind. They, and rubbing machines, waste energy because their random action involves repeated attack on some particles which have already been adequately disintegrated. Extrusion machines avoid that fault. Much ingenuity has been expended on devising methods for measuring to what extent leaf cells have been damaged by pulping so as to release protein. Among the methods suggested are: examination under the microscope, measurement of electrical conductivity of pulp, and measurement of the amount of chlorophyll in juice. However, the object of pulping is to release coagulable protein; it is not obvious that any of these methods for measuring cell damage would be preferable to a direct measurement of the coagulable (by heat or TCA) protein in juice.

Pulping is at present the most expensive phase of fodder fractionation, both in terms of energy used and cost of equipment. Several papers in Telek & Graham (1983) show that costs are being steadily diminished. Addy *et al.* (1983) claim that the job could be done with the expenditure of only 5% of the energy now used. If it becomes possible to approach that degree of efficiency, all existing economic forecasts would be invalidated.

Juice expression. Ideal conditions for expression are: a thin layer of pulp (p. 6) pressed by grids between which there is no lateral movement which would force fibre into the juice and waste energy. Double cone, horn-angle, and belt presses (Pirie, 1959*b*) approach the ideal to varied extents. Christensen (1984) describes an elaborate belt press which extracted 30% more proein than was extracted by a screw expeller. Conventional screw expellers have the basic defects, when working with unlubricated material such as leaf pulp, that compacted fibre is moved across metal surfaces, and that juice

Figure 12. A convenient unit for pressing juice from 10 to 20 kg lots of pulp. Leaf pulp is spread on the grid in the middle of the photograph. When the beam is lifted slightly by the handle on the extreme right, the hinged leg (which in the photograph is holding up the beam) is kicked forward. The beam is lowered slowly so as to avoid squeezing pulp out sideways. The amount of weight put on the beam depends on the physique of the operator. After 1 to 2 min the beam is lifted: the leg swings automatically into position to support the beam. A fresh lot of pulp is then slid under the platen and the pressed fibre is removed.

  To minimise contamination, aluminium, plastic or stainless steel are used for all parts coming into contact with juice.

is forced through a thick layer of fibre. It is well known (e.g. Davys *et al.*, 1969) that juice emerging after pulp has become tightly packed has only a third or less of the protein content of juice emerging with gentle presure. For the same reason, it is unwise, in small-scale work, to put several kilograms of pulp into a perforated container, with approximately equal height and width, and then

Figure 13. Illustration showing the principle of a press for separating juice and fibre on a large scale. Four springs, each with 5 mm compression under 200 kg load, support the wide-angle cone. They are not visible in the photograph. The rotating annulus has been lifted off the central shaft so as to show one arrangement of the paddles, on its under surface, which force the pulp towards the rim of the cone. The ideal paddle design has not yet been found. A pinion wheel is fastened to the upper surface of the annulus. Four of its teeth can be seen in the central hole. A chain connects that pinion wheel to the driving pinion on the left. An upper frame (not shown) carries the bearing for the shaft and four wheels which hold the annulus down. Juice goes through the perforated metal of the cone; fibre comes out all round the cone. Four revolutions per minute (rpm) is a convenient running speed.

apply pressure with a hydraulic cylinder or motor car jack. That procedure has recently been advocated.

Two arrangements used at Rothamsted partly circumvent these defects. In small-scale work, pulp is spread in a layer 1 to 2 cm thick on part of a grid, and a plate (40 × 25 cm) is pressed down on the pulp by a heavy, hinged lever (Figure 12). During the 1 to 2 min needed for juice expulsion, a new charge of pulp is arranged on another part of the grid. The lever is then lifted and the new charge is slid into position. With practice, 10 kg of pulp can be processed by one person in 10 to 12 min. For work on a larger scale, the principle of the screw expeller is retained but, instead of having a narrow-angle cone rotating within a perforated cylinder, an annulus rotates over a wide-angle, perforated cone (Figure 13). Angled paddles on the annulus, and ribs on the cone, force pulp, fed in through the centre of the annulus, into the narrowing gap between annulus and cone. There is still friction between metal surfaces and fibre, but pressure, applied by springs holding up the cone, is less than in a conventional screw expeller and it is applied more slowly. Fibre comes out as a shell, 2 or 3 mm thick, round the whole 2 m circumference of the unit, and it is drier than any fibre we have been able to make on a belt press.

Dilution? Obviously, protein extraction is increased if water is added to the pulp or during pulping. With a lush crop carrying some surface water as a result of dew or of washing dust from the leaf surfaces, adding more water is not likely to increase protein extraction to an extent which compensates for the extra cost of coagulating a more dilute juice. In several papers it is suggested that 'whey' should be used to increase protein extraction. It is hard to see the advantage. In the circumstances in which LP would be made, water is not expensive and there is no evidence that 'whey' mobilises more protein than would be mobilised by an equal volume of added water. The disadvantages of exposing LP to increased contact with the phenolic components of 'whey' have already been stated (p. 53).

Fibre. Leaf residue from which LP has been extracted is the main product of fodder fractionation. As has already been stressed (p. 132), the only sensible use for it is as a ruminant feed. If it is to be fed fresh or after ensiling, there is no advantage in pressing it very hard: the extra yield of dilute juice would probably not compensate for the extra cost of heavy equipment and power to drive it. The situation is different if the residue is to be dried as winter feed. Removal of water by pressing costs less (Pirie, 1977) than removal by evaporation when fuel is used to supply the latent heat, and the greater the removal of water by pressure, the more amenable the fibre becomes to air-drying in humid climates where it is at risk from fungal attack. Although it would complicate the process, it may then be worthwhile pressing in two stages: first gently to get protein-rich juice and then more heavily. It is easier to feed partly dewatered pulp into a press of the horn-angle type than to feed sloppy pulp into such a press.

Energy costs. Sceptics and advocates make many valid but antithetical points when arguing about the amount of energy that could be saved by fractionation. Latent heat of vaporisation need not be wasted. If unfractionated fodder were dried in a 'vapour-compression' unit, liquid water would be discharged and much of the latent heat would be recovered. But the cost of such units is of a different order from the cost of conventional driers. The energy in the fuel used in a conventional drier is applied directly to the crop, whereas mechanical energy is used in fractionation. Conversion of energy in fuel into electrical energy, or to mechanical energy in a diesel, is only 20 to 40% efficient. On the other hand, in a conventional drier the transfer to a crop of the energy of combustion is far from 100% efficient, and the energy 'wasted' by a diesel is discharged at a high enough temperature to be useful – for example, for coagulating juice or drying fibre (p. 52). Much more information is needed before the merits of fodder fractionation, as a means for conserving energy, can be accurately assessed. Meanwhile, we should heed the conclusion of an eminently practical paper by Bruhn *et al.* (1977) that the process '...may serve to delay the demise of the dehydration industry now operating on gas or liquid fuel.' Added weight will be given to that point when it is borne in mind that the crops most worth drying have the greatest water content, and that, to make maximum use of equipment, it is often necessary to work with crops with rain on them. LP extraction benefits rather than suffers from humidity.

Having discussed these uncertainties and imponderables, nothing more need be said about the economics of coagulation and the separation of LP from 'whey'. Such processes are familiar parts of industrial chemistry and their costs are well known.

Economic assessment: the value of the products

Uncertainties about the value of the products, though great, are not as great as uncertainties about the costs of production. The simplest case is a country where food and arable land are scarce and there is unemployment. No critic has produced a valid argument against the proposition that it would then be worthwhile fertilising more adequately the land used to produce fodder, feeding ruminants on the extracted residue and supplying people with the LP. With such a policy, LP would be one among several potentially valuable underexploited sources of protein (Pirie, 1984b). In circumstances such as these, costs of production matter less than the possibility of meeting a present need, and advocates do not envisage production soon on a scale which would supply more protein than is supplied, for example, by fish in the UK. This book is not concerned with sociology. Nevertheless, it is worth noting that several papers have commented on the part that LP production in a village could play in the general process of education and in countering the tendency of more alert villagers to 'drift' to town.

In well-fed industrialised countries, LP is not likely soon to become a common human food. It will, unfortunately, first come in, as *Spirulina* has, through health food shops on the basis of largely spurious alleged merits. LP will then probably be sold at an inflated price. Soluble, decolorised leaf protein may compete with protein extracted from legume seeds as a component of 'convenience' and similar foods. If that happens, the prices of four, rather than three, products will have to be assessed because only part of the protein in leaf juice can be separated in that form: most of it would be used as feed for pigs and poultry. However, large-scale production in wealthy countries will depend on the savings made when partly dewatered fibre is conserved as winter feed for ruminants. Jones (1981) estimated that fodder fractionation would increase annual profit in Aberdeenshire (UK) by £56 ha^{-1}. Schoney & McGuckin (1983) concluded that, by minimising risks from unpredictable weather, fractionation would be advantageous on farms of more than 40 ha. An estimate in Japan (Ikumo &

Takigawa, 1984) put LP equal to fish-meal and preferable to soya. This is probably equally true in Britain. It seems absurd that the British use so much land to grow wheat, for which the market is glutted, instead of using it to produce LP as a substitute for imported soya (Pirie, 1985*b*).

Uncertainty about the economics of novel processes is not unprecedented. Various forms of dried and powdered fish have been used for 2500 years and made commercially for 50 years. In spite of considerable favourable publicity, a recent review (Crisan, 1978) stated: 'It is difficult, if not impossible, to predict accurately what a marketable FPC (fish protein concentrate) will cost at this point in its development...'. Two further examples of possible sources of error in assessing the economics of LP production may be given. If a substantial fraction of the cattle in a country were being maintained in winter on conserved fibre residue, the associated pig and poultry industries could probably absorb all the LP as a *protein* source. But some economic arguments have given the LP extra value as a *xanthophyll* source. That is a market which might soon be flooded. By contrast, one gloomy forecast, instead of including 'whey' as an exploitable asset, included in the costs of production the cost of laying a 60 km pipeline to carry 'whey' to sea at the point nearest to the economists' institute!

From potentiality to actuality

As the idea of fodder fractionation spreads among those who feed animals and those who manufacture fodder, economic argument will be subjected to practical testing. It is clear from the extensive literature cited in this book that widespread interest has now been aroused. It seems reasonable to conclude that when proper extraction equipment has been designed, and experience has identified the circumstances in which fractionation is likely to be successful, fibre and LP will take their places among accepted animal feeding stuffs. The fate of the 'whey' is still uncertain.

Progress will be less straightforward with LP as a human food. A few enthusiasts, having the mistaken idea that LP is a complete food rather than a supplement, have asked for supplies for expeditions, or have proposed that extraction equipment should be sent to disaster areas. Research on a novel food should be done before, not during, a disaster. One sign of progress towards acceptance is that a novel (Mitchison, 1983) about the problems arising when a benevolent multinational company tries to solve the world's food problems, concludes that LP is the only solution to them. At the other extreme are sceptics who, seizing on every uncertainty or difficulty, condemn the whole idea in all circumstances. The middle course is slowly being followed. Sometimes, as in Ghana and Sri Lanka, the use of LP as a supplement fits smoothly into the local culinary pattern. Elsewhere, adoption of LP is unlikely until some person or group, with respected opinions, sets an example. Even then, as already pointed out (p. 105), advocates must be prepared for a few weeks delay before acceptance becomes general.

People may take a few weeks to accept a novel food: organisations take very much longer to accept a novel idea. Work on LP got little support 25 years ago because the 'official' attitude was that all would be well with the world's food supply if conventional methods were used a little more efficiently, and if a stockpile were maintained to mitigate intermittent famines. Ten to 15 years ago the 'climate of opinion' began to change. Some aspects of this change are summarised elsewhere (Pirie, 1975b). The basic principles of agriculture have enabled farmers to produce ever-increasing amounts of DM ha^{-1} year^{-1}, and to protect what is produced from predation and infection. Nevertheless, the idea is gaining ground that there are unexploited merits in many little-used species, and that even conventional species are not always exploited optimally. LP production is one of these novel forms of exploitation. It is not a panacea – it has, for example, little to offer in arid regions – but its use would benefit about a third of the world's population.

The technique of fodder fractionation is in its infancy. Economic assessment has been attempted on the basis of existing crops and existing equipment, with insufficient attention to the effects of selecting crops and designing equipment specifically for this purpose. Textbooks of biochemical engineering deal almost exclusively with microbial conversions and neglect the less wasteful process of separation. Conversions are often necessary, but separation will sometimes suffice. By making optimal use of all the products of a fractionation, processes that seem questionable in isolation can become viable. Hartley (1937) saw this clearly 50 years ago: 'Finally there is the social aspect of the problem. Hitherto agriculture and industry have represented divergent interests, the town and the village, the factory and the farm, the huge closely knit corporations wanting cheap food and raw materials, contrasted with the scattered, unorganised agricultural producers at the mercy of nature for their output. The modern development of factories in close touch with farmers in the processing of their goods should bring industry and agriculture closer together and help to establish a community of interest between them. What their joint efforts can accomplish will depend on the progress of knowledge in biochemistry and chemistry on the one hand, and in plant breeding and cultivation on the other. With the modern technique of genetics and the closer association of the farmer and the manufacturer, there is a fascinating prospect of new strains of plants that will yield the ideal products for industry almost to a standard specification. Then indeed we should have realised Bacon's ideal of commanding nature in action.'

That is one side of the picture. As knowledge of how to make LP spreads, nature can also be 'commanded in action' in the kitchen and village.

Appendix

Hilaire Marin Rouelle
1718–79

Rouelle's paper is now often referred to, but seems seldom to have been read. A free translation may therefore be useful. In modern French, *fécule* means starch. Rouelle used the word in more than one sense. It was possibly because of this that Fourcroy doubted his conclusions and suggested that Rouelle had simply ground the leaf to such a fine pulp that all of it passed through the cloth used for straining. This suggestion incensed Proust – the discoverer of leucine and an early contributor to our knowledge of citric acid. Proust (1803) wrote of the '…beautiful liquid velvet expressed from leaves', but he used *fécule* in as imprecise a manner as Rouelle. *Fécule*, and the various words for weights and volumes, are best left untranslated.

Observations on the Fécules or green parts of Plants, and on glutinous or vegeto-animal matter; by M. Rouelle, demonstrator in chemistry in the 'Jardin du Roi'

The *fécules* or green parts of plants were classed as resins by my late brother because of their solubility in all oily solvents and in alcohol. He defined these *fécules* as being composed of (1) green resinous colouring matter and (2) parenchyma or plant fibres, separated by pounding with a pestle; and he pointed out that when either the *fécules* or the plant juices were heated with oils or fats that dissolve the green part, some insoluble matter always remained which he thought, as I have just said, belonged to the parenchyma or fibrous portion of the plant.

I have given, in the *Journal de Médecine* of last March and in the *Avant Coureur*, etc., analyses of several *fécules* or green parts of plants. I have explained that *fécules* made from different plant families, after being dried, give, when analysed in a retort, the same

154

products as animal substances; this proves that the *fécules* or green-coloured parts of plants are not made of pure vegetable matter, because the analytical products of vegetable matter are not found in them, but on the contrary, those of animal matter.

When I wrote of this in the *Journal de Médecine* and when I said that the green *fécules* were not resins because the products of their analysis were quite different from those of all known resins', I did not think I needed to give a clearer explanation of the nature of this substance; I stated, however, that the presence in all vegetable matter of a substance completely similar to the glutinous matter of wheat could be demonstrated. I laid aside for a further paper the more precise description of this kind of substance, which is in fact composed of a pure resin which gives the green colour to all plants, and of this glutinous or vegeto-animal matter.

The glutinous matter, which is found in all the *fécules*, is usually more abundant than the resinous green part, the latter making up, in the plants that I have examined, only a fifth to a third of the *fécule*. I will confine myself for the moment to a single example, and indicate some others.

Fécule of hemlock

The required amount of hemlock, when it is almost in flower, is ground carefully in a marble mortar with a wooden pestle. The juice is strained through a well-stretched cloth, or a cloth filter, and heated to the point at which one can hold a finger in it for several minutes. The *fécule* separates out and part of it floats on top of the liquid; part sinks or remains suspended. The whole is put on a cloth filter, the liquid becomes clear; the sediment remains on the cloth and is carefully collected. This is the procedure usually followed to make these plant *fécules*.

Remarks

(I) By paying a little attention to what goes on in this process, it is clear that the greenest part of the *fécule* separates first. When the heat is increased, distinct and well-defined flakes or white dots appear below the *fécule* which floats on the surface. These observations suggest that there are two *fécules*.

(II) If juice is heated to the temperature of milk when drawn from a cow and then taken off the fire, the *fécule* which separates has a more beautiful green colour than the one in Remark I. If it is immediately poured into a cloth, the filtrate is still slightly green.

(III) If liquid separated as above is heated more strongly, a *fécule* separates which is slightly green. This second *fécule* contains more glutinous, or vegeto-animal matter, than the first although, as will be seen, that contains an abundance.

(IV) *Fécule* made in the usual way is put into earthenware or glazed vessels and diluted, using a wooden stirrer, by pouring into each vessel 8 or 9 *pintes* of water. After standing for 24 hours to allow the *fécule* to sediment, the water is decanted; this is repeated a second and third time, leaving it to stand for 24 hours each time. After the third wash, the *fécule* is put on a cloth mounted on a wooden frame so as to get out as much moisture as possible; the *fécule* and cloth are then put on a plaster tile to absorb most of the remaining water. The *fécule* becomes fairly solid, it is cut into little pieces and put to dry on paper placed on sieves.

(V) Several methods can be used to separate the two *fécules*; but some are difficult and others expensive. The easiest uses alcohol, which has no action on vegeto-animal matter but dissolves the green colouring matter. The dried *fécule* has only to be ground to a fine powder and digested with alcohol several times.

The alcohol dissolves the green part and leaves the vegeto-animal part; but the process takes a very long time because of the way the two substances adhere during drying; I have found that it is difficult to separate them completely.

(VI) The two substances are more easily separated if the above procedure is followed using fresh *fécule*. When it is ready to be dried, it is mixed carefully with alcohol, digested for 24 hours in a water bath or sand bath, and left to cool. The alcohol is then decanted or filtered. This digestion is repeated a second and third time. The three extracts are put into the tin boiler of a distilling apparatus; the alcohol passes over as a clear liquid which smells of hemlock. A soft, resinous substance, which clings to the fingers like terebinth resin, remains in the boiler. The alcohol that has been distilled off is used to re-extract the *fécule* and the process is repeated until the *fécule* no longer gives up any green colour. The quantity of green colouring matter varies a little; it depends first on the state and age of the plant, and secondly on whether the leaves or the stalks are taken.

(VII) The *fécule* remaining is a dirty greyish-white, and blackens on drying. It alone makes up three-quarters of the amount of matter used in the experiment; this is the glutinous or vegeto-animal substance, as its analysis will show.

(VIII) When 4 *onces* of this substance were slowly heated in a retort in a reverberatory furnace, a small amount of distillate and drops of volatile alkali came over first. When the heat was increased, the alkali became more concentrated, and finally became solid. At the same time, an oil passed over which floated on the alkaline fluid. This resembles the oil obtained from the vegeto-animal matter of flour and from the caseous part of milk.

The residue is fairly bulky. It has a fairly even texture because the pieces of the glutinous substance have softened and stuck together. It closely resembles the residue from the glutinous part of corn and the caseous part of milk. It weighs an *once* or more.

(IX) A few drops of distillate passed over when 2 *onces* of the green colouring matter were heated gently in a glass retort. When the heat was increased, an acid liquid came over and increased in strength, then a beautiful yellow oil which became darker as it thickened. The acid is very strong and resembles that obtained from wax. The oil is light and floats on the acid, as do most oils obtained from resins. Finally, the residue is fairly bulky and light and weighs 3 *gros* and 60 *grains*.

(X) Rosemary, from which the extractable part has been removed by repeated decoctions, is only, according to Boerhaave, the dross or skeleton of the plant. It still contains a small amount of crude oil which gives a little flame, but its ash does not give fixed alkali.

This learned doctor did not know that this exhausted rosemary gives fixed alkali and contains a green colouring matter which is soluble in fats, oils, resins and alcohol.

My brother, as I have already said, and as we have written – it was indeed printed long ago, and is in short known by everybody – was the first to demonstrate this green part in exhausted rosemary; but there are also in the rosemary left after extraction with water and alcohol, two substances, namely: (1) a very small quantity of glutinous or vegeto-animal matter, and (2) a substance that has a vegetable nature, insoluble in water and alcohol. The rosemary left after extraction with these two solvents still gives to a fairly marked degree some acid and some oil when distilled. Consequently, there remains in the rosemary a constituent or a substance which was hitherto unknown and which escapes the action of these two solvents. Later, I will describe this substance more fully.

References

Abo Bakr, T. M., El-Shemi, N. M. & Messallam, A. S. (1984). Isolation and chemical evaluation of protein from water hyacinth. *Qual. Plant. Plant Fds Hum. Nutr.*, **34**, 67.

Abu-Shakra, S. S., Phillips, D. A. & Huffaker, R. C. (1978). Nitrogen fixation and delayed leaf senescence in soybeans. *Science*, **199**, 973.

Addy, T. O., Whitney, L. F. & Chen, C. S. (1983). Mechanical parameters in leaf cell rupture for protein production. In Telek & Graham (1983), p. 490.

Ahmad, M. B., Akram, M., Jaffri, S. A., Iqbal, A. & Yusuf, A. (1980). Comparative nutritive value of berseem (*Trifolium alexandrinum*) leaf residue, cotton-seed cake (undec.) and maize gluten feed in fattening rations for Sahiwal calves. *J. agric. Res. (Lahore)*, **18**, 79.

Akeson, W. R. & Stahmann, M. A. (1965). Nutritive value of leaf protein concentrate, an *in vitro* digestion study. *J. agric. Fd Chem.*, **13**, 145.

Akinrele, I. A. (1963). The manufacture and utilisation of leaf protein. *J. W. Afr. Sci. Ass.*, **8**, 74.

Alcantara, P. F. & Lobos, A. D. (1981). Nutritional evaluation of water hyacinth (*Eichhornia crassipes*) as feed for swine. *Nat. Sci. Dev. Board Tech. J.*, **6**(2) 15.

Allison, F. E. (1973). *Soil organic matter and its role in crop production.* Elsevier, Amsterdam.

Allison, R. M. (1971). Factors influencing the availability of lysine in leaf protein. In Pirie (1971a), p. 78.

Allison, R. M. & Vartha, E. W. (1973). Yields of protein extracted from irrigated lucerne. *N.Z. J. exp. Agric.*, **1**, 35.

Allison, R. M., Laird, W. M. & Synge, R. L. M. (1973). Notes on a deamination method proposed for determining 'chemically available lysine' of proteins. *Br. J. Nutr.*, **29**, 51.

Ameenuddin, S., Bird, H. R., Pringle, D. J. & Sunde, M. L. (1983). Studies on the utilization of leaf protein concentrates as a protein source in poultry nutrition. *Poultry Sci.*, **62**, 505.

Ameenuddin, S., Bird, H. R., Pringle, D. J. & Sunde, M. L. (1984a). Studies on the utilization of leaf protein concentrates as a protein source in poultry nutrition. *Poultry Adviser*, **17**, 21.

Ameenuddin, S., Bird, H. R., Sunde, M. L. & Koegel, R. G. (1984b). Effect of added methionine and lysine on the performance of chicks fed different alfalfa protein concentrates. *Poultry Adviser*, **17**, 33.

Anderson, J. W. & Rowan, K. S. (1967). Extraction of soluble leaf enzymes with thiols and other reducing agents. *Phytochemistry*, **6**, 1047.

Anelli, G., Fiorentini, R., Massignan, L. & Galoppini, C. (1977). The poly-protein process: a new method for obtaining leaf protein concentrates. *J. Fd Sci.*, **42**, 1401.

Anonymous (1965). *Investigations into the production of high protein concentrate from leaves for inclusion in the diet of infants and children.* Sci. Res. Coun. Tech. Rep., Kingston, Jamaica.

Anonymous (1974). Alfalfa protein for human use. *Agric. Sci. Rev.*, 11(2), 55.

Anonymous (1984). Vitamin A and cancer. *Lancet*, **ii**, 325.

Arkcoll, D. B. (1969). Preservation of leaf protein by air drying. *J. Sci. Fd Agric.*, **20**, 600.

Arkcoll, D. B. (1971). Agronomic aspects of leaf protein production in Great Britain. In Pirie (1971a), p. 9.

Arkcoll, D. B. (1973a). The preservation and storage of leaf protein preparations. *J. Sci. Fd Agric.*, **24**, 437.

Arkcoll, D. B. (1973b). Effects of leaf protein 'whey' on soil. *A. Rep. Rothamsted exp. Stn*, 1972, 117.

Arkcoll, D. B. & Festenstein, G. N. (1971). A preliminary study of the agronomic factors affecting the yield of extractable leaf protein. *J. Sci. Fd Agric.*, **22**, 49.

Arkcoll, D. B. & Holden, M. (1973). Changes in chloroplast pigments during the preparation of leaf protein. *J. Sci. Fd Agric.*, **24**, 1217.

Bagchi, D. K. & Matai, S. (1976). Studies on the performance of tetrakali (*Phaseolus aureus* Linn.) as a leaf protein yielding crop in West Bengal. *J. Sci. Fd Agric.*, **27**, 1.

Bagchi, D. K. & Matai, S. (1978). Root yield and extracted protein yield from different varieties of radish (*Raphanus sativus*) under W. Bengal conditions. *Indian J. agric. Chem.*, **11**, 37.

Balasundaram, C. S. & Samuel, D. M. (1968). Incorporation of different leaf proteins in South Indian dishes. *Madras agric. J.*, **55**, 540.

Balasundaram, C. S., Krishnamoorthy, K. K., Chandramani, R., Balakrishnan, T. & Ramadoss, C. (1974a). The yield of leaf protein extracted by large scale processing of various crops. *Indian J. Home Sci.*, **8**, 6.

Balasundaram, C. S., Krishnamoorthy, K. K., Balakrishnan, T., Ramadoss, C. & Chandramani, R. (1974b). Screening plant species for leaf protein extraction. *Indian J. Home Sci.*, **8**, 1.

Balasundaram, C. S., Chandramani, R., Krishnamoorthy, K. K. & Balakrishnan, T. (1975). Optimum time of cutting for maximum yield of extractable protein from some fodder grasses. *Madras agric. J.*, **62**, 431.

Barbeau, W. E. & Kinsella, J. E. (1983). Factors affecting the binding of chlorogenic acid to fraction 1 leaf protein. *J. agric. Fd Chem.*, **31**, 993.

Barber, R. S., Braude, R. & Mitchell, K. G. (1959). Leaf protein in rations of growing pigs. *Proc. Nutr. Soc.*, **18**, iii.

Barnes, M. F. (1976). Growth of yeasts on spent lucerne whey and their effectiveness in scavenging residual protein. *N.Z. J. agric. Res.*, 19, 537.

Batra, U. R., Deshmukh, M. G. & Joshi, R. N. (1976). Factors affecting extractability of protein from green plants. *Indian J. Pl. Physiol.*, 19, 211.

Bawden, F. C. & Kleczkowski, A. (1945). Protein precipitation and virus inactivation by extracts of strawberry plants. *J. Pomol.*, 21, 2.

Bawden, F. C. & Pirie, N. W. (1938). A note on some protein constituents of normal tobacco and tomato leaves. *Br. J. exp. Path.*, 19, 264.

Bawden, F. C. & Pirie, N. W. (1944). The liberation of virus together with materials that inhibit its precipitation with antiserum, from the solid leaf residues of tomato plants suffering from bushy stunt. *Br. J. exp. Path.*, 25, 68.

Ben Aziz, A., Grossman, S., Budowski, P., Ascarelli, I. & Bondi, A. (1968). Antioxidant properties of lucerne extracts. *J. Sci. Fd Agric.*, 19, 605.

Ben Aziz, A., Grossman, S., Budowski, P. & Ascarelli, I. (1971). Enzymic oxidation of carotene and linoleate by alfalfa: properties of active fractions. *Phytochemistry*, 10, 1823.

Betschart, A. A. (1977). The incorporation of leaf protein concentrates and isolates in human diets. In Wilkins (1977), p. 83.

Betschart, A. A. & Kinsella, J. E. (1974a). Influence of storage on composition, amino acid content and solubility of soybean leaf protein concentrate, *J. agric. Fd Chem.*, 22, 116.

Betschart, A. A. & Kinsella, J. E. (1974b). Influence of storage on *in vitro* digestibility of soybean leaf protein concentrate. *J. agric. Fd Chem.*, 22, 672.

Bickoff, E. M. & Kohler, G. O. (1974). Preparation of edible protein of leafy green crops such as alfalfa. *US Pat.* 3 823 128.

Bickoff, E. M., Bevenue, A. & Williams, K. T. (1947). Alfalfa has a promising chemurgic future. *Chemurg. Dig.*, 6, 215.

Bickoff, E. M., Booth, A. N., de Fremery, D., Edwards, R. H., Knuckles B. E., Miller, R. E., Saunders, R. M. & Kohler, G. O. (1975). Nutritional evaluation of alfalfa leaf protein concentrate. In *Protein nutritional quality of foods and feeds*, ed. M. Friedman, p. 319. Dekker, New York.

Blaxter, K. & Waterlow, J. C. eds (1985). *Nutritional adaptation in man.* John Libbey, London.

Borhami, B. E. & El-Shazly, K. (1984). Utilisation of water hyacinth and berseem proteins, fibrous residues and wheys. In N. Singh (1984), p. 399.

Bosshard, H. (1972). Über die Anlagerung von Thioäthern an Chino·ie und Chinonimine in stark sauren Medien. *Helv. chim. Acta*, 55, 32.

Bourdon, D., Perez, J. M., Henry, Y. & Calmes, R. (1980). Valeur énergétique et azotée d'un concentré de protéines de luzerne, le PXi, et utilisation par le porc en croissance-finition. *Journées de la Rech. Porcine en France*, 227.

Boyd, C. E. (1968). Fresh-water plants: a potential source of protein. *Econ. Bot.*, 22, 359.

Boyd, C. E. (1971) Leaf protein from aquatic plants. In Pirie (1971a), p. 44.

Boyd, C. E. (1976). Accumulation of dry matter, nitrogen and phosphorus by cultivated water hyacinth. *Econ. Bot.*, 30, 51.

Braude, R., Jones, A. S. & Houseman, R. A. (1977). The utilization of the juice extracted from green crops. In Wilkins (1977), p. 47.

Bray, W. J. (1976). Leaf-protein concentrate as a source of vitamins. *Proc. Nutr. Soc.*, **35**, 6A.

Bray, W. J. (1983). The economics of green crop fractionation and leaf protein production. In Telek & Graham (1983), p. 633.

Bray, W. J., Humphries, C. & Ineritei, M. S. (1978). The use of solvents to decolourise leaf protein concentrate. *J. Sci. Fd Agric.*, **29**, 165.

Bris, E. J., Hibbits, A. G., Algeo, J. W., Wooden, G. R. & Batley, B. (1970). A comparison of rations with concentrated alfalfa extract and cane molasses as to volatile fatty acid production *in vitro* and digestibility *in vivo. J. Anim. Sci.*, **31**, 237.

Brot, N. & Weissbach, H. (1982). The biochemistry of methionine sulfoxide residues in proteins. *Trends in biochem. Sci.*, **7**, 137.

Brown, H. E. & Saldana, G. (1976). Protein yields from *Brassica carinata. J. Rio Grande Hort. Soc.*, **30**, 87.

Brown, H. E., Stein, E. R. & Saldana, G. (1975). Evaluation of *Brassica carinata* as a source of plant protein. *J. agric. Fd Chem.*, **23**, 545.

Bruhn, H. D., Straub, R. J. & Koegel, R. G. (1977). On farm forage protein – the potential and the means. Paper 3 at the A. Conf. Inst. agric. Engineers. London.

Bruhn, H. D., Straub, R. J. & Koegel, R. G. (1983). Equipment for plant juice protein extraction in developing countries. *Amer. Soc. Agric. Eng.*, paper 5515.

Bryant, A. M., Carruthers, V. R. & Trigg, T. E. (1983). Nutritive value of pressed herbage residues for lactating dairy cows. *N.Z. J. agric. Res.*, **26**, 79.

Bryant, M. & Fowden, L. (1959). Protein composition in relation to age of daffodil leaves. *Ann. Bot.*, **23**, 65.

Bubicz, M. & Jelinowska. A. (1983). Quality of products obtained in the extraction of protein from lucerne plants. I. Amino acid composition of protein concentrates from lucerne in relation to variety, cutting date and growth phase. *Roczniki Nauk Rolniczych*, **105**, 41.

Buchanan, A. R. (1969a). *In vivo* and *in vitro* methods of measuring nutritive value of leaf protein preparations. *Br. J. Nutr.*, **23**, 533.

Buchanan, A. R. (1969b). Effect of storage and lipid extraction on the properties of leaf protein. *J. Sci. Fd Agric.*, **20**, 359.

Buchanan, A. R. & Byers, M. (1969). Interference by cyanide with the measurements of papain hydrolysis. *J. Sci. Fd Agric.*, **20**, 364.

Butler, J. B. (1982). An investigation into some causes of the differences of protein expressibility from leaf pulps. *J. Sci. Fd Agric.*, **33**, 528.

Butler, J. B. & Pirie, N. W. (1981). An improved small scale unit for extracting leaf juice. *Expl. Agric.*, **17**, 39.

Butt, A. M., Ahmad, S. U. & Shah, F. H. (1972). Propagation of yeast on leaf protein concentrate by-products. *Pakist. J. scient. ind. Res.*, **15**, 208.

Byers, M. (1961). The extraction of protein from leaves of some plants growing in Ghana. *J. Sci. Fd Agric.*, **12**, 20.

Byers, M. (1967a). The *in vitro* hydrolysis of leaf proteins. I. The action of papain on protein extracted from the leaves of *Zea mays. J. Sci. Fd Agric.*, **18**, 28.

Byers, M. (1967b). II. The action of papain on protein concentrates extracted from leaves of different species. *J. Sci. Fd Agric.*, **18**, 33.

Byers, M. (1971a). The amino acid composition of some leaf protein preparations. In Pirie (1971a), p. 95.

Byers, M. (1971b). The amino acid composition and *in vitro* digestibility of some protein fractions from three species of leaves of various ages. *J. Sci. Fd Agric.*, **22**, 242.

Byers, M. (1975). Relationship between total N, total S and the S-containing amino acids in extracted leaf protein. *J. Sci. Fd Agric.*, **27**, 135.

Byers, M. (1983). Extracted leaf proteins: their amino acid composition and nutritional quality. In Telek & Graham (1983), p. 135.

Byers, M. & Jenkins, G. (1961). Effect of gibberellic acid on the extraction of protein from the leaves of spring vetches (*Vicia sativa* L.). *J. Sci. Fd Agric.*, **12**, 656.

Byers, M. & Sturrock, J. W. (1965). The yields of leaf protein extracted by large-scale processing of various crops. *J. Sci. Fd Agric.*, **16**, 341.

Byers, M., Green, S. H. & Pirie, N. W. (1965). The presentation of leaf protein on the table II. *Nutrition*, **19**, 63.

Carlsson, R. (1975). Selection of centrospermae and other species for production of leaf protein concentrates. Ph.D. thesis, University of Lund (Many copies of this thesis are in circulation.)

Carlsson, R. (1980). Quantity and quality of leaf protein concentrates from *Atriplex hortensis* L., *Chenopodium quinoa* Willd. and *Amaranthus caudatus* L., grown in southern Sweden. *Acta Agric. Scand.*, **30**, 418.

Carlsson, R. (1983). Leaf protein concentrate from plant sources in temperate climates. In Telek & Graham (1983), p. 52.

Carlsson, R., Jokl, L. & Santos, R. C. (1984). Effects of processing conditions on the chemical composition and nutritive value of leaf protein concentrates from tropical legumes and from leaves of forest trees. In N. Singh (1984), p. 221.

Carpenter, K. J., Duckworth, J. & Ellinger, G. M. (1952). The supplementary protein value of a by-product from grass processing. *Br. J. Nutr.*, **6**, xii.

Carr, J. R. & Pearson, G. (1974). Nutritive values of lucerne leaf-protein concentrate and lupin-seed meal as protein supplements to barley diets for growing pigs. *N.Z. Soc. Anim. Prod.*, **34**, 95.

Carr, J. R. & Pearson, G. (1976). Photosensitisation, growth performance and carcass measurements of pigs fed diets containing commercially prepared lucerne leaf-protein concentrate. *N.Z. J. exp. Agric.*, **4**, 45.

Carruthers, I. B. & Pirie, N. W. (1975). The yields of extracted protein, and of residual fibre, from potato haulm taken as a by-product. *Biotechnol. Bioeng.*, **17**, 1775.

Carton, O. & Maguire, M. F. (1983). The nutritive value of preserved grass juice for growing pigs. *Irish J. agric. Res.*, **22**, 95.

Casselman, T. W., Green, V. E., Allen, R. J. & Thomas, F. H. (1965). *Mechanical dewatering of forage crops. Agric. exp. Stn Univ. Florida Tech. Bull.*, **694**.

Chakrabarti, S., Bagchi, D. K. & Matai, S. (1984). Use of by-product leaves of some vegetable crops as source of protein based on some nutritional parameters. In N. Singh (1984), p. 233.

Chalmers, M. I. & Synge, R. L. M. (1954). The digestion of protein and nitrogenous compounds in ruminants. *Adv. Protein Chem.*, **9**, 93.

Chanda, S. (1983). Whey utilization for penicillin production: present status and future prospects. In Roy (1983), p. 527.

Chanda, S., Chakrabarti, S. & Bagchi, D. K. (1980). Propagation of yeast in whey, a by-product from leaf protein production plant. *Current Sci.*, **49**, 793.

Chandramani, R., Balasundaram, C. S., Krishnamoorthy, K. K. & Balakrishnan, T. (1975a). Effect of nitrogen and the frequency of cutting in guinea grass (*Panicum maximum* L.). *Madras agric. J.*, **62**, 155.

Chandramani, R., Krishnamoorthy, K. K., Balasundaram, C. S. & Balakrishnan, T. (1975b). Optimum time of cutting for obtaining maximum yield of extractable protein from fenugreek (*Trigonella foenum-graecum*) varieties. *Madras agric. J.*, **62**, 230.

Chayen, I. H. (1959). Protein production by the impulse process. *Engineering, London*, **188**, 307.

Chayen, I. H., Smith, R. S., Tristram, G. R., Thirkell, D. & Webb, T. (1961). The isolation of leaf components. *J. Sci. Fd Agric.*, **12**, 502.

Cheeke, P. R. (1974). Nutritional evaluation of alfalfa protein concentrate with rats, swine and rabbits. In *Proc. 12th tech. alfalfa conf. USDA*, p. 76, USDA, Washington.

Cheeke, P. R. & Garman, G. R. (1974). Influence of dietary protein and sulfur amino acid levels on the toxicity of *Senecio jacobaea* (Tansy ragwort) in rats. *Nutr. Rep. Int.*, **9**, 193.

Cheeke, P. R., Kinzell, J. H., de Fremery, D. & Kohler, G. O. (1977). Freeze dried and commercially-prepared alfalfa protein concentrate evaluation with rats and swine. *J. Anim. Sci.*, **44**, 772.

Cheeke, P. R., Telek, L., Carlsson, R. & Evans, J. J. (1980). Nutritional evaluation of leaf protein concentrates prepared from selected tropical plants. *Nutr. Rep. Int.*, **22**, 717.

Cheeseman, G. C. (1977). The chemical composition of forage juice and its preservation. In Wilkins (1977), p. 39.

Chen, I. & Mitchell, H. L. (1973). Trypsin inhibitors in plants. *Phytochemistry*, **12**, 327.

Chibnall, A. C. (1939). *Protein metabolism in the plant*. Yale Univ. Press, New Haven.

Chibnall, A. C., Rees, M. W. & Lugg, J. W. H. (1963). The amino acid composition of leaf proteins. *J. Sci. Fd Agric.*, **14**, 234.

Christensen, I. L. H. (1984). Methods of increasing yield of food and feed grade LPC in industrial scale. In N. Singh (1984), p. 143.

Clare, N. T. (1952). Photosensitization in diseases of domestic animals. *Commonw. Bur. Anim. Hlth Rev. Ser.*, **3**.

Clarke, E. M. W. & Ellinger, G. M. (1967). Fractionation of plant material. 2. Amino acid composition of some fractions obtained from broad bean plant (*Vicia faba* L.) and chromatographic differentiation of hydroxyproline isomers. *J. Sci. Fd Agric.*, **18**, 536.

Cohen, M., Ginoza, W., Dorner, R. W., Hudson, W. R. & Wildman, S. G. (1956). Solubility and color characteristics of leaf proteins prepared in air and nitrogen. *Science*, **124**, 1081.

Coleman, G. S. (1983). Hydrolysis of fraction 1 leaf protein and casein by rumen entodiniomorphid protozoa. *J. appl. Bact.*, **55**, 111.

Colker, D. A., Eskew, R. K. & Aceto, N. C. (1948). Preparation of vegetable leaf meals. *USDA Tech. Bull.*, **958**, 53.

Commonwealth Science Council (1981). *Second review meeting on management of water hyacinth*. Colombo, June 1981. Marlborough House, Pall Mall, London, SWIY 5HX.

Connell, J. (1975). The prospects for green crop fractionation. *Span*, **18**, 103.

Connell, J. & Foxell, P. R. (1976). Green crop fractionation, the products and their utilization by cattle, pigs and poultry. *Bienn. Rev. natn. Inst. Res. Dairying*, 1976, 21.

Connell, J. & Houseman, R. A. (1977). The utilisation by ruminants of the pressed green crops from fractionation machinery. In Wilkins (1977), p. 51

Cooke, B. C. (1978). A study of the relationship between β carotene and fertility problems in dairy cows. *Animal Production*, **26**, 356.

Costes, C. ed. (1981). *Protéines foliaires et alimentation*. Gauthier-Villars, Paris.

Cowey, C. B., Pope, J. A., Adron, J. W. & Blair, A. (1971). Studies on the nutrition of marine flatfish. Growth of the plaice (*Pleuronectes platessa*) on diets containing proteins derived from plants and other sources. *Mar. Biol.*, **10**, 145.

Cowlishaw, S. J., Eyles, D. E., Raymond, W. F. & Tilley, J. M. A. (1956a). Nutritive value of leaf protein concentrates. I. Effect of addition of cholesterol and amino-acids. *J. Sci. Fd Agric.*, **7**, 768.

Cowlishsaw, S. J., Eyles, D. E., Raymond, W. F. & Tilley, J. M. A. (1956b). II. Effects of processing methods. *J. Sci. Fd Agric.*, **7**, 775.

Crisan, E. V. (1978). Fish protein concentrate (FPC). In *Encyclopedia of food science*, eds M. S. Peterson & A. H. Johnson, p. 266. AVI Publishing Co., Westport, Conn., USA.

Crook, E. M. (1946). The extraction of nitrogenous materials from green leaves. *Biochem. J.*, **40**, 197.

Crook, E. M. & Holden, M. (1948). Some factors affecting the extraction of nitrogenous materials from leaves of various species. *Biochem. J.*, **43**, 181.

Dakore, H. G. & Mungikar, A. M. (1985). Quality of silage prepared from fibrous residue left after leaf protein extraction from lucerne (*Medicago sativa* L.). *Indian J. agric. Chem.*, **17**, 205.

Darwin, C. (1868). *Variation of plants and animals under domestication*. Murray, London.

Davies, A. M. C., Newby, V. K. & Synge, R. L. M. (1978). Bound quinic acid as a measure of coupling of leaf and sunflower-seed proteins with chlorogenic acid congeners: loss of availability of lysine. *J. Sci. Fd Agric.*, **29**, 33.

Davies, M., Evans, W. C. & Parr, W. H. (1952). Biological values and digestibilities of some grasses, and protein preparations from young and mature species, by the Thomas-Mitchell method, using rats. *Biochem. J.* **52**, xxiii.

Davies, R., Laird, W. M. & Synge, R. L. M. (1975). Hydrogenation as an approach to study of reactions of oxidizing polyphenols with plant proteins. *Phytochemistry*, **14**, 1591.

Davies, W. L. (1926). The proteins of green forage plants. 1. The proteins of some leguminous plants. *J. agric. Sci. Camb.*, **16**, 280.

Davys, G. (1981). Letter to the editor. *Appropriate Technol.* 8(3), 12.

Davys, M. N. G. & Pirie, N. W. (1960). Protein from leaves by bulk extraction. *Engineering, London*, **190**, 274.

Davys, M. N. G. & Pirie, N. W. (1963). Batch production of protein from leaves. *J. agric. Engng Res.*, **8**, 70.

Davys, M. N. G. & Pirie, N. W. (1965). A belt press for separating juices from fibrous pulps. *J. agric. Engng Res.*, **10**, 142.

Davys, M. N. G. & Pirie, N. W. (1969). A laboratory-scale pulper for leafy plant material. *Biotechnol. Bioeng.*, **11**, 517.

Davys, M. N. G., Pirie, N. W. & Street, G. (1969). A laboratory-scale press for extracting juice from leaf pulp. *Biotechnol. Bioeng.*, **11**, 528.

Dayrell, M. S. & Vieira, E. C. (1977). Leaf protein concentrate of the cactacea *Pereskia aculeata* Mill. *Nutr. Rep. Int.*, **15**, 529 & 539.

Deepchand, K. (1984). Leaf protein from cane tops and leaves – a study of extraction methods. In N. Singh (1984), p. 21.

de Fremery, D., Bickoff, E. M. & Kohler, G. O. (1972). PRO-XAN process: air drying of alfalfa leaf protein concentrate. *J. agric. Fd Chem.*, **20**, 1155.

de Fremery, D., Miller, R. E., Edwards, R. H., Knuckles, B. E., Bickoff, E. M. & Kohler, G. O. (1973). Centrifugal separation of white and green protein fractions from alfalfa juice following controlled heating. *J. agric. Fd Chem.*, **21**, 866.

de Fremery, D., Edwards, R. H., Miller, R. E., Knuckles, B. E., Bickoff, E. M. & Kohler, G. O. (1974). Composition and uses of PRO-XAN and pressed residue. In *Proc. 12th tech. alfalfa conf. USDA*, p. 73. USDA, Washington.

De Jong, D. W. (1984). Manipulation of agronomic and process factors for maximizing leaf protein extractability. In N. Singh (1984), p. 9.

Derbyshire, J. C., Gordon, C. H., Holdren, R. D. & Menear, J. R. (1969). Evaluation of dewatering and wilting as moisture reduction methods for hay-crop silage. *Agron. J.*, **61**, 928.

Deshmukh, M. G. & Joshi, R. N. (1969). Leaf protein from some leguminous plants. *Sci. Cult.*, **35**, 629.

Deshmukh, M. G. & Joshi, R. N. (1973). Effect of rhizobial inoculation on the extraction of protein from the leaves of cowpea (*Vigna sinensis* L. Savi ex Hassk.). *Indian J. agric. Sci.*, **43**, 539.

Deshmukh, M. G., Gore, S. B., Mungikar, A. M. & Joshi, R. N. (1974). The yields of leaf protein from various short-duration crops. *J. Sci. Fd Agric.*, **25**, 717.

Deshmukh, V. R., Deshmukh, M. G., Basole, G. R. & Joshi, R. N. (1984). Short duration crops for leaf protein extraction in Marathwada. In N. Singh (1984), p. 33.

Dev, D. V. & Joshi, R. N. (1969). Extraction of protein from some plants of Aurangabad. *J. biol. Sci.*, **12**, 15.

Dev, D. V., Batra, U. R. & Joshi, R. N. (1974). The yields of extracted leaf protein from lucerne (*Medicago sativa* L.). *J. Sci. Fd Agric.*, **25**, 725.

Devadas, R. P. (1981). Appropriate technology with reference to infant weaning foods. *Proc. 1st Household Nutr. Approp. Technol. Conf., Colombo, Sri Lanka*, p. 199.

Devadas, R. P., Kupputhai, A. & Sebastian, S. (1978). Effect of leaf protein on nitrogen retention in pre-school children. *Indian. J. Nutr. Dietet.*, **15**, 107.

Devadas, R. P., Vijayalakshmi, P. & Vijaya, S. (1984). Studies on nutritional trials with pre-school children with low cost leaf protein supplements. In N. Singh (1984), p. 311.

Devi, A. V., Rao, N. A. N. & Vijayaraghavan, P. K. (1965). Isolation and composition of leaf protein from certain species of Indian flora. *J. Sci. Fd Agric.*, **16**, 116.

Donnelly, P. E., McDonald, R. M., Mills, R. A., Ritchie, J. M., Swan, J. E., Trigg, T. E. & Bryant, A. M. (1980). Protein from pasture. *Ruakura Farm Conf. Proc.*, **43**, Hamilton, New Zealand.

Donnelly, P. E., McDonald, R. M. & Rattray, P. V. (1983). Protein extraction from pasture: the effect of crop species and of a reducing agent on the quality of the extracted protein. *J. Sci. Fd Agric.*, **34**, 828.

Donnelly, P. E. & Rattray, P. V. (1983). Protein extraction from pasture: the nutritional availability of methionine, cystine and lysine in leaf protein concentrates. *J. Sci. Fd Agric.*, **34**, 839.

Doraiswamy, T. R., Singh, N. & Daniel, V. A. (1969). Effects of supplementing ragi (*Eleusine coracana*) diets with lysine or leaf protein on the growth and nitrogen metabolism of children. *Br. J. Nutr.*, **23**, 737.

Dove, M. R. (1983). Theories of swidden agriculture, and the political economy of ignorance. *Agroforestry Systems*, **1**, 85.

Duckworth, J. & Woodham, A. A. (1961). Leaf protein concentrates. I. Effect of source of raw material and method of drying on protein value for chicks and rats. *J. Sci. Fd Agric.*, **12**, 5.

Duckworth, J., Hepburn, W. R. & Woodham, A. A. (1961). Leaf protein concentrates. II. The value of a commercially dried product for newly-weaned pigs. *J. Sci. Fd Agric.*, **12**, 16.

Dumont, A. G. & Boyce, D. S. (1976). Leaf protein production and use on the farm: an economic study. *J. Br. Grassld Soc.*, **31**, 153.

Dutrow, G. F. (1971). *Economic implications of silage sycamore. USDA Forest Serv. Res. Pap.*, SO 66.

Dye, M., Medlock, O. C. & Christ, J. W. (1927). The association of vitamin A with greenness in plant tissue. I. The relative vitamin A content of head and leaf lettuce. *J. biol. Chem.*, **74**, 95.

Dykyjova, D. (1971). Productivity and solar energy conversion in reed-swamp stands in comparison with outdoor mass cultures of algae in the temperate climate of central Europe. *Photosynthetica*, **5**, 329.

Edwards, R. H., Miller, R. E., de Fremery, D., Knuckles, B. E., Bickoff, E. M. & Kohler, G. O. (1975). Pilot-plant production of an edible white fraction leaf protein concentrate from alfalfa. *J. agric. Fd Chem.*, **23**, 620.

Eggum, B. O. & Christensen, K. D. (1975). Influence of tannin on protein utilization in feedstuffs with special reference to barley. In *Breeding for seed protein improvement*, p. 135. International Atomic Energy Agency, Vienna.

Eichenberger, W. & Grob, E. C. (1965). Beiträge zur Chemie der pflanzlichen Plastiden. *Helv. chim. Acta*, **48**, 1094.

Eidrigevich, V. E., Kinsburgskii, Z. S., Pidorenko, L. D., Naumenko, V. I. &

Pasechnik, G. I. (1978). Nutritive value of the fractionation products of lucerne. *Vest. Selsk. Nauki, (Moscow)*, **6**, 95.

Ellinger, G. M. (1978). A chemical approach to the nutritional availability of methionine in food proteins. *Ann. Nutr. Aliment.*, **32**, 281.

Elliott, K. & Knight, J. eds. (1972). *Lipids, malnutrition and the developing brain.* Ciba Foundation symposium, Elsevier, Amsterdam.

Enochian, R. V., Kohler, G. O., Edwards, R. H., Kuzmicky, D. D. & Vosloh, C. J. (1983). Economics of producing LPC for feed with the Pro-Xan process. In Telek & Graham (1983), p. 525.

Ereky, K. (1927). Process for the manufacture and preservation of green fodder pulp or other plant pulp and of dry products made therefrom. *Br. Pat.* 270 629.

Fafunso, M. & Bassir, O. (1976). Effects of age and season on yield of crude and extractable proteins from some edible plants. *Expl Agric.*, **12**, 249.

Fafunso, M. & Byers, M. (1977). Effect of pre-press treatments of vegetation on the quality of the extracted leaf protein. *J. Sci. Fd Agric.*, **28**, 375.

Fafunso, M. & Oke, O. L. (1977). Leaf protein from different cassava varieties. *Nutr. Rep. Int.*, **14**, 629.

Fantozzi, P. & Sensidoni, A. (1983). Protein extraction from tobacco leaves: technological, nutritional and agronomical aspects. *Qual. Plant., Plant Fds Hum. Nutr.*, **32**, 351.

Feeny, P. P. (1969). Inhibitory effect of oak leaf tannins on the hydrolysis of proteins by trypsin. *Phytochemistry*, **8**, 2119.

Ferrando, R. & Spais, A. (1966). Valeur alimentaire des protéines extraites de la luzerne. *Proc. 7th Congr. Int. Nutr.*, **5**, 276.

Festenstein, G. N. (1961). Extraction of proteins from green leaves. *J. Sci. Fd Agric.*, **12**, 305.

Festenstein, G. N. (1972). Water-soluble carbohydrates in extracts from large-scale preparations of leaf protein. *J. Sci. Fd Agric.*, **23**, 1409.

Festenstein, G. N. (1976). Carbohydrates associated with leaf protein. *J. Sci. Fd Agric.*, **27**, 849.

Find Your Feet (1985). Untitled. *Appropriate Technol.* **12**(1), 20.

Finot, P. A. & Mauron, J. (1972). Le blocage de la lysine par la réaction de Maillard. II. Propriétés chimiques des dérivés N-(désoxy-l-D-fructosyl-) et N-(désoxy-l-D-lactulosyl-l) de la lysine. *Helv. chim. Acta*, **55**, 1153.

Fiorentini, R., Galoppini, C. & Lepidi, A. A. (1984). Progress in wet green crop fractionation in Italy. In N. Singh (1984), p. 417.

FAO (1964). *The state of food and agriculture.* FAO, Rome.

FAO (1969). *Provisional indicative world plan for agricultural development.* FAO, Rome.

FAO (1971). *Production yearbook.* FAO, Rome.

FAO (1973). *Energy and protein requirements. Tech. Rep. Ser.*, **522**, FAO, Rome.

Foot, A. S. (1974). Lucerne juice for pigs. *Pig Fmg*, **22**(9), 71.

Ford Foundation (1959). *Report on India's food crisis and steps to meet it.* Ford Foundation, New Delhi.

Foreman, F. W. (1938). Observations on the proteins of pasturage. Phosphorus and protoplasm. *J. agric. Sci. Camb.*, **28**, 135.

Fourcroy, A. F. (1789). Mémoire sur l'existence de la matière albumineuse dans les végétaux. *Annls Chim.*, **3**, 252.

Foxell, P. R. (1977). The separation and preservation of leaf-protein concentrate for animal feeds. In Wilkins (1977), p. 97.

Franzen, K. L. & Kinsella, J. E. (1976). Functional properties of succinylated and acetylated leaf protein. *J. agric. Fd Chem.*, **24**, 914.

Fujihara, T. & Ohshima, M. (1980). The effect of formaldehyde and formic acid treatment on the utilization of fibrous residue silage made from Ladino clover in sheep. *J. Jap. Soc. Grassld Sci.*, **26**, 191.

Fujihara, T. & Ohshima, M. (1984). Utilization of dried fibrous residue left after the extraction of legume leaf protein in sheep. *J. Jap. Soc. Grassld Sci.* **30**, 284.

Gangawane, L. V. & Nehemiah, K. M. A. (1980). Preparation of rhizobial inoculants using deproteinized alfalfa liquor rhizosphere effect. *Geobios*, **7**, 43.

Garcha, J. S., Kawatra, B. L., Wagle, D. S. & Bhatia, I. S. (1970). Studies on extraction and isolation of leaf protein of various crops grown in the Punjab. *J. Res. Punjab agric. Univ.*, **7**, 211.

Garcha, J. S., Kawatra, B. L. & Wagle, D. S. (1971). Nutritional evaluation of leaf proteins and the effect of their supplementation to wheat flour by rat feeding. *J. Fd Sci. Technol.*, **8**, 23.

Gardner, H. W. (1979). Lipid hydroperoxidation with proteins and amino acids: a review. *J. agric. Fd Chem.*, **27**, 220.

Gastineau, C. (1974). The French work. In *Proc. 12th Tech. alfalfa conf. USDA*, p. 123. USDA, Washington.

Gastineau, C. (1976). Advertising material from France-Luzerne. France-Luzerne, 173 Avenue de Alliés, 51 Chalons S/Marne, France.

Gastineau, I. & de Mathan, O. (1984). Leaf protein extraction technology and research in Champagne: achievements and researches by France-Luzerne. In N. Singh (1984), p. 433.

Gerloff, E. D., Lima, I. H. & Stahmann, M. A. (1965). Amino acid composition of leaf protein concentrates. *J. agric. Fd Chem.*, **13**, 139.

Gersovitz, M., Motil, K., Munro, H. N., Scrimshaw, N. S. & Young, V. R. (1982). Human protein requirements: assessment of the adequacy of the current Recommended Dietary Allowance for dietary protein for elderly men and women. *Am. J. Clin. Nutr.*, **35**, 6.

Ghosh, J. J. (1967). Leaf protein concentrates: problems and prospects in the control of protein malnutrition. *Trans. Bose Res. Inst. (Calcutta)*, **30**, 215.

Gibson, A. J. & Wallace, G. M. (1980). Changes in leaf protein concentrates on storage. *N.Z. J. Sci.*, **23**, 59.

Gjøen, A. U. & Njaa, L. R. (1977). Methionine sulphoxide as a source of sulphur-containing amino acids for the young rat. *Br. J. Nutr.*, **37**, 93.

Glencross, R. G., Festenstein, G. N. & King, H. G. C. (1972). Separation and determination of isoflavones in the protein concentrate from red clover leaves. *J. Sci. Fd Agric.*, **23**, 371.

Gonzalez, G. & Alzueta, C. (1984). Yield, chemical composition and nutritive value of the pressed pea vines (*Pisum sativum* L.) after LPC extraction. In N. Singh (1984), p. 355.

Gonzalez, G., Richelet, A. & Ocio, E. (1977). Efectos de la modificacion del pH en el jugo de plantas, sobre la cantidad y la composicion quimico-bromatologica de los correspondientes concentrados proteicos vegetales (CVP). *Rev. Nutr. Anim.*, **15**, 15.

Gonzalez, O. N., Dimaunahan, L. B. & Banyon, E. A. (1968). Extraction of protein from the leaves of some local plants. *Philipp. J. Sci.*, **97**, 17.

Goodall, C. (1936). Improvements relating to the treatment of grass and other vegetable substances. *Br. Pat.* 457 789.

Goodall, M. (1950). New high protein feed produced from sugar beet tops. *Br. Sug. Beet Rev.*, **19**, 54.

Gopalam, A. & Anthinarayanan, R. (1984). Protein fractionation studies in tobacco leaves. In N. Singh (1984), p. 41.

Gordon, C. H., Holdren, R. D. & Derbyshire, J. C. (1969). Field losses in harvesting wilted forage. *Agron. J.*, **61**, 924.

Gore, S. B. & Joshi, R. N. (1972). The exploitation of weeds for leaf protein. In *Symposium on tropical ecology*, eds. P. M. Golley & S. B. Golley, p. 137. Athens, USA.

Gore, S. B. & Joshi, R. N. (1976a). Effect of fertiliser and frequency of cutting on the extraction of protein from *Sesbania*. *Indian J. Agron.*, **21**, 39.

Gore, S. B. & Joshi, R. N. (1976b). A note on the effect of simazine on the yields of dry matter and crude protein, and on the extractability of protein from hybrid napier grass. *Indian J. Agron.*, **21**, 491.

Gore, S. B., Mungikar, A. M. & Joshi, R. N. (1974). The yield of extracted protein from hybrid napier grass. *J. Sci. Fd Agric.*, **25**, 1149.

Greenhalgh, J. F. D. & Reid, G. W. (1975). Mechanical processing of wet roughage. *Proc. Nutr. Soc.*, **34**, 74A.

Griffiths, T. W. & Maguire, M. F. eds (1983). *Forage protein conservation and utilisation.* Commission of the European Communities, Dublin.

Grover, A., Arora, S. O., Atreja, P. P. & Chopra, R. C. (1980). Effect of feeding berseem juice on growth rate of calves. *Indian J. Dairy Sci.*, **33**, 278.

Guha, B. C. (1960). Leaf protein as a human food. *Lancet*, i, 705.

Hageman, R. H. & Waygood, E. R. (1959). Methods for the extraction of enzymes from cereal leaves with especial reference to the triosephosphate dehydrogenases. *Pl. Physiol.*, **34**, 396.

Hanczakowski, P. (1983). Leaf protein research in Poland. In Telek & Graham (1983), p. 795.

Hanczakowski, P. & Makuch, M. (1980). The composition and nutritive value of protein concentrates from potato haulms. *Potato Res.*, **23**, 1.

Hanczakowski, P. & Skraba, B. (1984). Effect of different precipitating agents on quality of leaf protein concentrate from lucerne. *Animal Feed Sci. Tech.*, **12**, 11.

Hanczakowski, P., Skraba, B. & Mlodkowski, M. (1981). Nutritive value of leaf protein concentrate from potato haulm for rats and chicks. *Animal Food Sci. Tech.*, **6**, 413.

Harendranath, R. & Singh, N. (1980). Effect of pre-press hot water treatment in processing of vegetation for leaf protein. *Current Sci.* **49**, 152.

Hartley, H. (1937). Agriculture as a potential source of raw materials for industry. *J. Text. Inst.*, **28**, 151.

Hartman, G. H., Akeson, W. R. & Stahmann, M. A. (1967). Leaf protein concentrate by spray-drying. *J. agric. Fd Chem.*, **15**, 74.

Heath, S. B. (1977). The production of leaf protein concentrates from forage crops. In *Plant proteins*, ed. G. Norton, p. 171. Butterworth, London.

Heath, S. B. & King, M. W. (1977). The production of crops for green crop fractionation. In Wilkins (1977), p. 9.

Hegsted, M. & Linkswiler, H. M. (1980). Protein quality of high and low saponin alfalfa protein concentrate. *J. Sci. Fd Agric.*, **31**, 777.

Heidelberger, M., Kendall, F. E. & Scherp, H. W. (1936). The specific polysaccharides of types I, II and III pneumococcus. A revision of methods and data. *J. exp. Med.*, **64**, 559.

Henry, K. M. & Ford, J. E. (1965). The nutritive value of leaf protein concentrates determined in biological tests with rats and by microbiological methods. *J. Sci. Fd Agric.*, **16**, 425.

Herndon, J. H., Steinberg, D., Uhlendorf, B. W. & Fales, H. M. (1969). Refsum disease (HAP): characterisation of enzyme defect in cell culture. *J. clin. Invest.*, **48**, 1017.

Herrick, A. M. & Brown, C. L. (1967). Silage sycamore. *Agric. Sci. Rev.*, **4**, 8.

Hicks, D. R. & Crookston, R. K. (1976). Defoliation boosts corn yield. *Crops and Soils*, **29**, 12.

Hill-Cottingham, D. G. & Lloyd-Jones, C. P. (1979). Translocation of nitrogenous compounds in plants. In *Nitrogen assimilation of plants*, eds E. J. Hewitt & C. V. Cutting, p. 397. Academic Press, London.

Hinde, R. & Smith, D. C. (1975). Role of photosynthesis in the nutrition of the mollusc *Elysia viridis*. *Biol. J. Linnean Soc.*, **7**, 161.

Holden, M. (1945). Acid-producing mechanisms in minced leaves. *Biochem. J.*, **39**, 172.

Holden, M. (1952). The fractionation and enzymic breakdown of some phosphorus compounds in leaf tissue. *Biochem. J.*, **51**, 433.

Holden, M. (1974). Chlorophyll degradation products in leaf protein preparations. *J. Sci. Fd Agric.*, **25**, 1427.

Holden, M. & Tracey, M. V. (1948). The effect of fertilizers on the levels of nitrogen, phosphorus, protease, and pectase in healthy tobacco leaves. *Biochem. J.*, **43**, 147.

Holden, M. & Tracey, M. V. (1950). A study of enzymes that can break down tobacco-leaf components. 4. Mammalian pancreatic and salivary enzymes. *Biochem. J.*, **47**, 421.

Holl, W. & Hampp, R. (1975). Lead in plants. *Residue Rev.*, **54**, 79.

Hollo, J. & Koch, L. (1971). Commercial production in Hungary. In Pirie (1971a), p. 63.

Horigome, T. (1977). Nutritional studies on fractionated cytoplasmic and chloroplastic proteins from leaves of oats and ladino clover. *Jap. J. Zootech.*, **48**, 267.

Horigome, T. (1983). Leaf protein research and development in Japan. In Telek & Graham (1983), p. 731.

Horigome, T. (1984). Nutritional value of leaf protein concentrates prepared under different conditions. In N. Singh (1984), p. 239.

Horigome, T. & Kandatsu, M. (1964). Studies on the nutritive value of grass proteins: XII. Digestibilities of isolated grass proteins. *J. agric. Chem. Soc. Japan*, **38**, 121.

Horigome, T. & Uchida, S. (1980). An observation on the nutritional quality of leaf protein in connection with its methionine content. *Jap. J. Zootech. Sci.*, **51**, 429.

Horn, M. J., Lichtenstein, H. & Womack, M. (1968). Availability of amino acids: a methionine-fructose compound and its availability to microorganisms and rats. *J. agric. Fd Chem.*, **16**, 741.

Houseman, R. A. & Connell, J. (1976). The utilization of the products of green-crop fractionation by pigs and ruminants. *Proc. Nutr. Soc.*, **35**, 213.

Hove, E. L. & Bailey, R. W. (1975). Towards a leaf protein concentrate industry in New Zealand. *N.Z. J. exp. Agric.*, **3**, 193.

Hove, E. L., Lohrey, E., Urs, M. K. & Allison, R. M. (1974). The effect of lucerne-protein concentrate in the diet on growth, reproduction and body composition of rats. *Br. J. Nutr.*, **31**, 147.

Huang, K. H., Tao, M. C., Boulet, M., Riel, R. R., Julien, J. P. & Brisson, G. J. (1971). A process for the preparation of leaf protein concentrates based on the treatment of leaf juices with polar solvents. *Can. Inst. Fd Technol.*, **4**(3), 85.

Hudson, B. J. F. & Karis, I. G. (1973). Aspects of vegetable structural lipids: I. The lipids of leaf protein concentrate. *J. Sci. Fd Agric.*, **24**, 1541.

Hudson, B. J. F. & Karis, I. G. (1974). Aspects of vegetable structural lipids: II. The effect of crop maturity on leaf lipids. *J. Sci. Fd Agric.*, **25**, 1491.

Hudson, B. J. F. & Mahgoub, S. E. O. (1980). Naturally-occurring antioxidants in leaf lipids. *J. Sci. Fd Agric.*, **31**, 646.

Hudson, B. J. F. & Warwick, M. J. (1977). Lipid stabilisation in leaf protein concentrates from ryegrass. *J. Sci. Fd Agric.*, **28**, 259.

Hudson, J. P. (1975). Weeds as crops. In *Proc. 12th Br. weed control conf.*, p. 333. Br. Crop Protection Coun., London.

Hudson, J. P. (1976). Food crops for the future. *J. R. Soc. Arts*, **124**, 572.

Hume, E. M. (1921). Comparison of the growth-promoting properties for guinea-pigs of certain diets, consisting of natural foodstuffs. *Biochem. J.*, **15**, 30.

Hume, E. M. & Krebs, H. A. (1949). Vitamin A requirement of human adults. *Med. Res. Coun. Spec. Rep.*, **264**.

Humphries, C. (1980). Trypsin inhibitors in leaf protein concentrates. *J. Sci. Fd Agric.*, **31**, 1225.

Humphries, C. & Bray, W. J. (1979). Preparation of white leaf protein concentrate using a polyanionic flocculant. *J. Sci. Fd Agric.*, **30**, 171.

Hussain, A., Ullah, M. & Ahmad, B. (1968). Studies on the potentials of leaf proteins for the preparation of concentrates from various leaf-wastes in West Pakistan. *Pakist. J. agric. Res.*, **6**, 110.

Hutchinson, R. W. (1978). The dormancy of seed potatoes. 1. The effect of time of haulm destruction and harvesting. *Potato Res.*, **21**, 257.

Ikumo, H. & Takigawa, A. (1984). Economic evaluation of experimentally domestically prepared leaf protein concentrate as a poultry feed. *Jap. J. Zootech. Sci.*, **55**, 475.

Ivins, J. D. (1973). Increasing productivity: crop physiology and nutrition. *Phil. Trans. R. Soc. Lond.*, B, **267**, 81.

Jadhav, B. & Joshi, R. N. (1982). The extractability of protein from lucerne (*Medicago sativa* L.) which had been left drying after dipping in water and alkali. *Annls Arid Zone*, **21**, 15.

Jadhav, B., Tekale, N. S. & Joshi, R. N. (1979). Green-manure crops as a source of leaf protein. *Indian J. agric. Sci.*, **49**, 371.

James, A. T. & Larbey, D. M. (1976). New food technologies and their role in the world. In *People and food tomorrow*, eds D. Hollingsworth & E. Morse, p. 115. Applied Science Publishers, London.

Jelinowska, A. & Magnuszewska, K. (1981). Increment dynamics of the above-ground plant matter and leaf area in several alfalfa cultivars in connection with protein yields. *Biol. Oceny Odmian*, **9**, 209.

Jennings, A. C. & Watt, W. B. (1967). Fractionation of plant material. I. Extraction of proteins and nucleic acids from plant tissues and isolation of protein fractions containing hydroxyproline from broad bean (*Vicia faba* L.) leaves. *J. Sci. Fd Agric.*, **18**, 527.

Jennings, A. C., Pusztai, A., Synge, R. L. M. & Watt, W. B. (1968). Fractionation of plant material. III. Two schemes for chemical fractionation of fresh leaves, having special applicability for isolation of the bulk protein. *J. Sci. Fd Agric.*, **19**, 203.

Jokl, L. & Carlsson, R. (1984). Nutritive value of leaf protein concentrates from tropical legumes and from leaves of forest trees. *Nutr. Rep. Int.* **30**, 87.

Jollans, J. L. ed. (1981). *Grassland in the British economy*. Centre for Agricultural Strategy, Reading, UK.

Jones, A. S. (1981). Potential change in animal output from grassland: production from fractionated forage. In Jollans (1981), p. 496.

Jones, A. S. (1983). The effect of mechanical processing of grass on the nutritive value of forage for ruminants and the degradability in the rumen. In Griffiths & Maguire (1983), p. 47.

Jones, W. T. & Mangan, J. L. (1976). Large-scale isolation of fraction 1 leaf protein (18 S) from lucerne (*Medicago sativa* L.). *J. agric. Sci. Camb.*, **86**, 495.

Jönsson, A. G. (1962). Studies in the utilization of some agricultural wastes and by-products by various microbial processes. *K. LantbrHögsk. Annlr.*, **28**, 235.

Joshi, R. N. (1971). The yields of leaf protein that can be extracted from crops of Aurangabad. In Pirie (1971a).

Joshi, R. N. (1983). Leaf protein research in India. In Telek & Graham p. 673.

Joshi, R. N. & Mungikar, A. M. (1983). Efficiency of protein extraction from the fresh crop of lucerne. *Proc. Ind. Acad. Sci.*, **92**, 35.

Joshi, R. N., Savangikar, V. A. & Patunkar, B. W. (1983). Production of leaf protein in an Indian village – prospects and problems. In Roy (1983), p. 480.

Joshi, R. N, Savangikar, V. A. & Patunkar, B. W. (1984). The Bidkin green crop fractionation process. *Indian bot. Reptr.*, **3**, 136.

Kamalanathan, G. & Devadas, R. P. (1971). Acceptability of food preparations containing leaf protein concentrates. In Pirie (1971a).

Kanev, S., Boncheva, I., Georgieva, L. & Iovchev, N. (1976). Tests of a protein concentrate from lucerne for fattening pigs and poultry. Quoted from *Nutr. Abstr. Rev.*, **46**, 282.

Kannangara, C. G. & Stumpf, P. K. (1972). Fat metabolism in higher plants. I. The biosynthesis of polyunsaturated fatty acids by isolated spinach chloroplasts. *Arch. Biochim. Biophys.*, **148**, 414.

Kasture, M. N. & Mungikar, A. M. (1984). Intercropping short duration crops for leaf protein production. In N. Singh (1984), p. 49.

Kasture, M. N., Dakore, H. G., Shahane, J. & Mungikar, A. M. (1984). Conservation of fibrous residues left after leaf protein extraction by ensiling. In N. Singh (1984), p. 363.

Kawatra, B. L., Garcha, J. S. & Wagle, D. S. (1974). Effect of supplementation of leaf protein extracted from berseem (*Trifolium alexandrinum*) to wheat flour diet. *J. Fd Sci. Technol.*, **11**, 241.

Kayama, R. (1984). Pilot-plant for production of leaf protein concentrates in Japan. In N. Singh (1984), p. 439.

Kehr, W. R., Ogden, R. L. & Satterlee (1979). An alfalfa protein concentrate from four cultivars at three growth stages. *Agron. J.*, **71**, 272.

Kempton, T. J., Nolan, J. V. & Leng, R. A. (1977). Principles for the use of non-protein nitrogen and by-pass protein in diets of ruminants. *Wld Anim. Rev.*, **22**, 2.

Khandelwal, S. R., Baukhandi, S. S. & Andhale, M. S. (1984). Role of lactobacilli in preservation of leaf protein concentrate. In N. Singh (1984), p. 155.

Kiesel, A., Belozersky, A., Agatov, P., Biwshich, N. & Pawlowa, M. (1934). Vergleichende Untersuchungen über Organeiweiss von Pflanzen. *Z. physiol. Chem.*, **226**, 73.

Knuckles, B. E., de Fremery, D., Bickoff, E. M. & Kohler, G. O. (1975). Soluble protein from alfalfa juice by membrane filtration. *J. agric. Fd Chem.*, **23**, 209.

Knuckles, B. E., de Fremery, D. & Kohler, G. O. (1976). Coumestrol content of fractions obtained during wet processing of alfalfa. *J. agric. Fd Chem.*, **24**, 1177.

Knuckles, B. E., Edwards, R. H., Kohler, G. O. & Whitney, L. F. (1980*a*). Flocculants in the separation of green and soluble white protein fractions from alfalfa. *J. agric. Fd Chem.*, **28**, 32.

Knuckles, B. E., Edwards, R. H., Miller, R. E. & Kohler, G. O. (1980*b*). Pilot scale ultrafiltration of clarified alfalfa juice. *J. Fd Sci.*, **45**, 730.

Koch, L. (1974). Producing protein concentrates from green plants. *Växtodling (Plant Husbandry)*, **28**, 129.

Koch, L. (1983). The Vepex process. In Telek & Graham (1983), p. 601.

Koegel, R. G. & Bruhn, H. D. (1977). An analysis of the requirements for expression of plant juice. In Wilkins (1977), p. 23.

Korniewicz, A., Mazanowska, A. & Gwara, T. (1980). Effect of a protein concentrate from lucerne juice on productivity of hens and quality and yield of eggs. *Rocz. Nauk. Zootech., Monographie i Rozprawy*, **16**, 63.

Kotake, Y. & Knoop, F. (1911). Ueber einen krystallisierten Eiweisskörper aus dem Milchsäfte der *Antiaris toxicaria*. *Z. physiol. Chem.*, **75**, 488.

Kümmerlin, R. R. (1984). Production of single-cell protein and ethanol from deproteinized plant extracts. *Agro Sur*, **12**, 93.

Kumprecht, I., Gasnárek, Z., Prokop, V. & Jakobe, P. (1984). The effect of lucerne protein-vitamin concentrate on some nutritional and biochemical factors of chick broilers. Živoč. Výr., **29**, 545.

Kung, S. D., Saunder, J. A., Tso, D. A., Vaughan, M., Womack, M., Staples, R. C. & Beecher, G. R. (1980). Tobacco as a potential food source and smoke material: nutritional evaluation of tobacco leaf protein. *J. Fd Sci.*, **45**, 320.

Lahiry, N. L., Satterlee, L. D., Hsu, H. W. & Wallace, G. W. (1977). Characterization of the chlorogenic acid binding fraction in leaf protein concentration. *J. Fd Sci.* **42**, 83.

Laird, W. M., Mbadiwe, E. I. & Synge, R. L. M. (1976). A simplified procedure for fractionating plant mateials. *J. Sci. Fd Agric.*, **27**, 127.

Lala, V. R. & Reddy, V. (1970). Absorption of β carotene from green leafy vegetables in undernourished children. *Am. J. clin. Med.*, **23**, 110.

Lawes, J. B. (1864). On the chemistry of the feeding of animals for the production of meat and manure. *Farmers Gazette*, Dublin, Ireland.

Lawes, J. B. (1885). Sugar as a food for stock. *J. R. agric. Soc.*, **21**, 81.

Lea, C. H. & Parr, L. J. (1961). Some observations on the oxidative deterioration of the lipids of crude leaf protein. *J. Sci. Fd Agric.*, **12**, 785.

Levere, T. H. (1984). Dr Thomas Beddoes (1750–1808): science and medicine in politics and society. *Br. J. Hist. Sci.*, **17**, 187.

Levin, D. A. (1971). Plant phenolics: an ecological perspective. *Am. Nat.*, **105**, 157.

Levin, D. A. (1976). The chemical defenses of plants to pathogens and herbivores. *A. Rev. Ecol. Syst.*, **7**, 121.

Lexander, K., Carlsson, R., Schalen, V., Simonsson, A. & Lundborg, T. (1970). Quantities and qualities of leaf protein concentrates from wild species and crop species grown under controlled conditions. *Ann. appl. Biol.*, **66**, 193.

Li Sui Fong, J. C. (1982). Potential uses of cane tops and leaves – a preliminary study. *Int. Sugar J.*, **84**, 5.

Lima, I. H., Richardson, T. & Stahmann, M. A. (1965). Fatty acids in some leaf protein concentrates. *J. agric. Fd Chem.*, **13**, 143.

Livingstone, A. L., Mohler, G. O. & Kuzmicky, D. D. (1980). Comparison of carotenoid storage stability in alfalfa leaf protein (Pro-Xan) and dehydrated leaf meals. *J. agric. Fd Chem.*, **28**, 652.

Lohrey, E., Tapper, B. & Hove, E. L. (1974). Photosensitization of albino rats fed on lucerne-protein concentrates. *Br. J. Nutr.*, **31**, 159.

Lu, C. D., Jorgensen, N. A. & Barrington, G. P. (1980). Intake, digestibility, and rate of passage of silages and hays from wet fractionation of alfalfa. *J. Dairy Sci.*, **63**, 2051.

Lu, C. D., Jorgensen, N. A., Straub, R. J. & Koegel, R. G. (1981). Quality of alfalfa protein concentrate with changes in processing conditions during coagulation. *J. Dairy Sci.*, **64**, 1561.

Lu, C. D., Jorgensen, N. A. & Amundson, C. H. (1982). Ruminal degradation and intestinal absorption of alfalfa protein concentrate by sheep. *J. Anim. Sci.*, **54**, 1251.

Lu, C. D., Jorgensen, N. A. & Slater, L. D. (1983*a*). Quantitative studies of amino acid flow in the digestive tract of sheep fed alfalfa protein concentrates. *J. Nutr.*, **113**, 2390.

Lu, C. D., Jorgensen, N. A. & Slater, L. D. (1983*b*). Comparative study of site and extent of nutrient digestion in lactating dairy cows fed alfalfa protein concentrate or soybean meal. *J. Dairy Sci. Suppl. 1*, **66**, 187.

Lugg, J. W. H. (1932). The application of phospho-18-tungstic acid (Folin's reagent) to the colorimetric determination of cysteine, cystine and related substances. II. The determination of sulphydryl compounds and disulphides already existing in solution. *Biochem. J.*, **26**, 2160.

Lugg, J. W. H. (1939). The representativeness of extracted samples and the efficiency of extraction of protein from the fresh leaves of plants; and some partial analyses of the whole proteins of leaves. *Biochem. J.*, **33**, 110.

Lundborg, T. (1980*a*). Fractionation of leaf proteins by differential centrifugation and gel filtration. *Physiol. Plant.*, **48**, 175.

Lundborg, T. (1980*b*). Fractionation by centrifugation of leaf proteins in press juices from *Brassica* and other species as a function of pH. *Physiol. Plant.*, **48**, 186.

Lundborg, T. (1980*c*). Fractionation by centrifugation of leaf proteins from *Brassica napus, Brassica oleracea, Helianthus annuus,* and *Atriplex hortensis* as a function of pH and temperature. *Physiol. Plant.*, **48**, 251.

Lundborg, T. (1980*d*). Fractionation by centrifugation of leaf proteins from varieties and cultivars of *Helianthus annuus* and *Helianthus debilis* as a function of pH and temperature. *Physiol. Plant.*, **48**, 317.

Lyon, C. K. & Kohler, G. O. (1981). Leaf protein concentrates from leucaena leaves. *Leucaena Res. Rep.*, **2**, 81.

Lyon, C. K., Knowles, P. F. & Kohler, G. O. (1983). Evaluation of *Brassica* species as leaf sources for extending the processing season of a leaf protein concentrate plant. *J. Sci. Fd Agric.*, **34**, 849.

Magoon, M. L. (1972). March towards self-sufficiency in animal food. *Indian Fmg*, **22**, 25.

Maguire, M. F. & Brookes, I. M. (1972). Grass juice as a liquid feed. *Res. Rep. An Foras Talúntais. Anim. Prod.*, **38**.

Maguire, M. F. & Brookes, I. M. (1973). The effects of juice extraction on the composition and yield of grass crops for dehydration. In *Proc. 1st int. congr. green crop drying*, p. 346. Ass. Green Crop Driers, Oxford.

Maguire, M. F., Finn, P. & Carton, O. (1983). The utilisation of protein extracts obtained by fractionation of grass crops. In Griffiths & Maguire (1983), p. 35.

Mahadeviah, S. & Singh, N. (1968). Leaf protein from the green tops of *Cichorium intybus* L. (chicory). *Indian J. exp. Bio.*, **6**, 193.

Maliwal, B. P. (1983). *In vitro* methods to assess the nutritive value of leaf protein concentrate. *J. agric. Fd Chem.*, **31**, 315.

Marchaim, U., Birk, Y., Dovrat, A. & Berman, T. (1972). Lucerne saponins as inhibitors of cotton seed germination: their effect on diffusion of oxygen through seed coats. *J. exp. Bot.*, **23**, 302.

Martin, C. (1986). *All grass is flesh*.

Mason, J. B., Ahlers, T. A., Henderson, C., Shorr, I. J. & Tabatabai, H. (1985). Identifying nutritional considerations in planning a rural development project in Haiti. *Ecol. Fd Nutr.*, **18**, 1.

Masters-Thomas, A., Bailes, J., Billimoria, J. D., Clemens, M. E., Gibberd, F. B. & Page, R. G. (1980). Heredopathia atactica polyneuritiformis (Refsum's disease) 1. Clinical features and dietary management. 2. Estimation of phytanic acid in foods. *J. Human Nutr.*, **34**, 245 & 251.

Matai, S. (1976). Protein from water weeds. In *Aquatic weeds in South East Asia*, ed. C. K. Varshney & J. Rzoska, p. 369. Junk, The Hague.

Matai, S. (1984). *Leaf Protein Newsletter*, 7–9. (These and subsequent numbers can be had free from S. Matai at Indian Statistical Institute, 203 Barrackpore Trunk Road, Calcutta 700 035, India.)

Matai, S. & Bagchi, D. K. (1974). Some promising legumes for leaf protein extraction. *Sci. Cult.*, **40**, 34.

Matai, S., Bagchi, D. K. & Roy Chowdhury, S. (1971). Leaf protein from some plants in West Bengal. *Sci. Engng.*, **24**, 102.

Matai, S., Bagchi, D. K. & Chanda, S. (1973). Optimal seed rate and fertilizer dose for maximum yield of extracted protein from the leaves of mustard (*Brassica nigra* Koch) and turnip (*Brassica rapa* L.). *Indian J. agric. Sci.*, **43**, 165.

Matai, S., Bagchi, D. K. & Chanda, S. (1976). Effects of seed rate, nitrogen level and leaf age on the yield of extracted protein from five different crops in West Bengal. *J. Sci. Fd Agric.*, **27**, 736.

Matai, S., Banerjee, A. & Bagchi, D. (1980). Effect of nitrogen on fodder and extracted protein yield of *bajra* in Gangetic alluvial soil. *Indian J. agric. Chem.*, **13**, 81.

Mathismoen, P. (1974). Stord twin screw presses: design and application on alfalfa. In *Proc. 12th tech. alfalfa conf. USDA*, p. 135. USDA, Washington.

Mauron, J. (1970a). Le comportement chimique des protéines lors de la préparation des aliments et ses incidences biologiques. *J. int. Vitaminol.*, **40**, 209.

Mauron, J. (1970b). Nutritional evaluation of proteins by enzymatic methods. In *Evaluation of novel protein products*, ed. A. E. Bender, R. Kihlberg, B. Löfquist & L. Munk, p. 211. Pergamon, Oxford.

McClure, J. W. (1970). Secondary constituents of aquatic angiosperms. In *Phytochemical phylogeny*, ed. J. B. Harborne, p. 233. Academic Press, London.

McDonald, R. M., Mills, R. A. & Donnelly, P. E. (1986). Practical aspects of commercial leaf-protein production. In Tasaki (1986), p. 29.

McDougall, V. D. (1980). Support energy and green crop fractionation in the United Kingdom. *Agric. Systems*, **5**, 251.

McLeay, L. M., Kokich, D. C., Hockey, H-U. & Trigg, T. E. (1982). Motility of the reticulum and rumen of sheep given juice-extracted pasture. *Br. J. Nutr.*, **47**, 79.

McLeod, M. N. (1974). Plant tannins – their role in forage quality. *Nutr. Abstr. Rev.*, **44**, 804.

McNeil, P. L. & Smith, D. C. (1982). The green hydra symbiosis. IV. Entry of symbionts into digestive cells. *Proc. R. Soc., B*, **216**, 1.

Mead, J. F. & Alfin-Slater, R. b. (1966). Toxic substances present in food fats. *Natn. Acad. Sci./Natn. Res. Coun. Publ.*, **1354**, p. 238.

Mehrotra, O. N., Mohan, M. & Srivastava, G. P. (1978). Note on the extractability and chemical composition of leaf protein from *Trianthema monogyna* L. *Indian J. agric. Chem.*, **11**, 81.

Merodio, C., Martin, M. & Sabater, B. (1983). Improved separation of green and soluble leaf proteins by pH shift. *J. agric. Fd Chem.*, **31**, 957.

Milic, B. L. (1972). Lucerne tannins. I. Content and composition during growth. *J. Sci. Fd Agric.*, **23**, 1151.

Milic, B. L. & Stojanovic, S. (1972). Lucerne tannins. III. Metabolic fate of lucerne tannins in mice. *J. Sci. Fd Agric.*, **23**, 1163.

Miller, D. S. (1965). Some nutritional problems in the utilization of nonconventional proteins for human feeding. *Recent Adv. Fd Sci.*, **3**, 125.

Miller, D. S. & Samuel, P. D. (1970). Effects of addition of sulphur compounds to the diet on utilisation of protein in young growing rats. *J. Sci. Fd Agric.*, **21**, 616.

Miller, R. E., Edwards, R. H., Lazar, M. E., Bickoff, E. M. & Kohler, G. O. (1972). PRO-XAN process: air drying alfalfa leaf protein concentrate. *J. agric. Fd Chem.*, **20**, 1151.

Miller, R. E., de Fremery, D., Bickoff, E. M. & Kohler, G. O. (1975). Soluble protein concentrate from alfalfa by low-temperature acid precipitation. *J. agric. Fd Chem.*, **23**, 1177.

Miller, S. A. (1983). Synthetic foods: technical, cultural and legal issues. In *Chemistry and world food supplies*, ed. L. W. Shemilt, p. 319. Pergamon Press, Oxford.

Mills, R. A. (1984). Performance of a forage harvester/dejuicer for leaf protein extraction. *Trans. Am. Soc. Agric. Eng.*, **27**, 1634.

Milovanov, L. V., Perel'dik, D. N., Kazakov, E. N. & Glazov, E. M. (1984). Green protein concentrate for polar foxes and mink (quoted from *Nutr. Abs. Rev.* (B), **55**, 475, 1985).

Mitchison, N. (1983). *Not by bread alone*. Marion Boyars, London.

Mitscher, L. A. (1975). Antimicrobial agents from higher plants. *Recent Adv. Phytochem.*, **9**, 243.

Mohan, M. & Srivastava, G. P. (1981). Studies on the extractability and chemical composition of leaf proteins from certain trees. *J. Fd Sci. Technol.*, **18**, 48.

Mokady, S. & Zimmermann, C. (1966). The effect of different liquid extractants used with impulse-rendered lucerne paste on the nutritional and calorigenic properties of the proteins. *Proc. 7th Int. Congr. Nutr.*, **5**, 279.

Monsod, G. G. (1976). *The versatility and economics of water hyacinth*. Philip. Coun. agric. Resources Res., Fisheries Forum, Manila.

Moore, P. D. (1976). Chemical antagonism in plant communities. *Nature*, **259**, 447.

Mori, T., Tsuji, T., Sugiura, M., Taniguchi, M. & Kobayashi, T. (1984). Effect of drying methods on quality of leaf protein concentrate. *Agric. biol. Chem.*, **48**, 1901.

Morris, P. & Hall, D. O. (1982). The inherent stability of *Chenopodium quinoa* chloroplasts. *Plant Sci. Letters*, **25**, 353.

Morris, T. R. (1977). Leaf protein concentrate for non-ruminant animals. In Wilkins (1977), p. 67.

Morrison, J. E. & Pirie, N. W. (1960). The presentation of leaf protein on the table. *Nutrition*, **14**, 7.

Morrison, J. E. & Pirie, N. W. (1961). The large-scale production of protein from leaf extracts. *J. Sci. Fd Agric.*, **12**, 1.

Mudgett, R. E., Rajogopalan, K. & Rosenau, J. R. (1980). Single cell protein recovery from alfalfa process wastes. *Trans. Am. Soc. Agric. Eng.*, **23**, 1590.

Mungikar, A. M. & Joshi, R. N. (1976). Studies on the ensilage of the residues left after extraction of leaf protein from lucerne and hybrid napier grass. *Indian J. Nutr. Dietet.*, **13**, 39.

Mungikar, A. M., Tekale, N. S. & Joshi, R. N. (1976a). The yields of leaf protein and fibre that can be obtained from fractionation of berseem (*Trifolium alexandrinum* L.). *Indian J. Nutr. Dietet.*, **13**, 114.

Mungikar, A. M., Batra, U. R., Tekale, N. S. & Joshi, R. N. (1976b). Effects of nitrogen fertilisation on the yields of extracted protein from some crops. *Expl Agric.*, **12**, 353.

Munshi, S. K., Wagle, D. S. & Thapar, V. K. (1974). Nutritional evaluation of leaf protein isolates treated under different drying temperatures. *Labdev J. Sci. Technol. ind.*, **12B**, 35.

Murai, M., Makino, K., Manda, T. & Takano, N. (1980). Feed value of leaf protein concentrates (LPC). *Bull. Nat. Grassld Res. Inst.*, **17**, 106.

Murai, M., Makino, K. & Manda, T. (1982). Studies on feed value of leaf protein concentrate (LPC). 3. The effects of crude LPC in the diet on growth of chicks. *Bull. Nat. Grassld Res. Inst.* **23**, 87.

Murai, M., Makino. K. & Manda. T. (1984). Studies on feed value of leaf protein concentrate (LPC). 4. The effect of giving feeds containing soluble fractions of crude LPC by extracting with organic solvents on growth and body composition in the chicks. *Bull. Nat. Grassld Res. Inst.*, **28**, 65.

Myer, R. O., Cheeke, P. R. & Kennick, W. H. (1975). Utilization of alfalfa protein concentrate by swine. *J. Anim. Sci.*, **40**, 885.

Nagy, S. & Nordby, H. E. (1983). Lipids in leaf protein concentrates. In Telek & Graham (1983), p. 268.

Nanda, C. L., Ternouth, J. H. & Kondos, A. C. (1977). Evaluation of the nutritive value of plant protein concentrates. *J. Sci. Fd Agric.*, **28**, 1075.

Näsi, M. (1983a). Leaf protein production from energy willow leaves. *J. sci. agric. Soc. Finland*, **55**, 155.

Näsi, M. (1983b). Preservation of grass juice and wet leaf protein concentrate for animal feeds. *J. sci. agric. Soc. Finland*, **55**, 465.

Näsi, M. & Kiiskinen, T. (1985). Leaf protein from green pulse crops and nutritive value of legume protein concentrates for poultry. *J. agric. Sci. Finland*, **57**, 117.

Nass, M. M. K. (1969). Uptake of isolated chloroplasts by mammalian cells. *Science*, **165**, 1128.

National Economic Development Office (1974). *UK farming and the Common Market: grass and grass products*. NEDO, London.

Naumenko, V., Tarasenko, A. & Kinsburgskii, Z. (1975). Juice from lucerne in feeds for young pigs. Quoted from *Nutr. Abstr. Rev.*, **45**, 580.

Naumenko, V., Morozova, A. & Kinsburgskii, Z. (1977). Lucerne juice in diets for pigs. Quoted from *Nutr. Abstr. Rev.*, **47**, 341.

Nazir, M. & Shah, F. H. (1966). Extractability of proteins from various leaves. *Pakist. J. scient. ind. Res.*, **9**, 235.

Nazir, M. & Shah, F. H. (1985). Studies on Persian clover (*Trifolium resupinatum*). Part II Effect of climatic changes and harvesting intervals on the yield of extractable protein and other fractions of Persian clover. *Qual. Plant., Plant Fds Hum. Nutr.*, **35**, 57.

Nedorizescu, M. (1972). Production of some fodder meals made from different forest products. Quoted from *Nutr. Abstr. Rev.*, 1974, **44**, 234.

Newby, V. K., Sablon, R-M., Synge, R. L. M., Casteele, K. V. & Van Sumere, C. F. (1980). Free and bound phenolic acids of lucerne (*Medicago sativa* CV Europe). *Phytochemistry.*, **19**, 651.

Njaa, L. R. (1962). Some problems related to detection of methionine sulfoxide in protein hydrolysates. *Acta chem. scand.*, **16**, 1359.

Njaa, L. R. & Aksnes, A. (1982). The nitrogen-sparing effect of methionine sulphoxide and some other sulphur-containing amino acids. *Br. J. Nutr.*, **48**, 565.

Nørgaard Pedersen, E. J. (1983). Ensiling of pressed crops. In Griffiths & Maguire (1983), p. 259.

Nørgaard Pedersen, E. J., Witt, N., Mortensen, J. & Sorensen, C. (1980). Fractionation of green crops and preservation of pressed crops and juice. 1. Ensiling of pressed crops. *Tids. Planteavl*, **84**, 265.

Nowakowski, T. Z. & Byers, M. (1972). Effects of nitrogen and potassium fertilisers on contents of carbohydrates and free amino acids in Italian ryegrass. II. Changes in the composition of the non-protein nitrogen fraction and the distribution of individual amino acids. *J. Sci. Fd Agric.*, **23**, 1313.

Nugent, J. H. A., Jones, W. T., Jordan, D. J. & Mangan, J. L. (1983). Rates of proteolysis in the rumen of the soluble proteins casein, fraction 1(18s) protein, bovine serum albumin and bovine submaxillary mucoprotein. *Br. J. Nutr.*, **50**, 357.

Nugent, J. H. A. & Mangan, J. L. (1981). Characteristics of the rumen proteolysis of fraction 1(18s) leaf protein from lucerne (*Medicago sativa* L.). *Br. J. Nutr.*, **46**, 39.

Oelshlegel, F. J., Schroeder, J. R. & Stahmann, M. A. (1969). Potential for protein concentrates from alfalfa and waste green plant material. *J. agric. Fd Chem.*, **17**, 791.

Ogino, C., Cowey, C. B. & Chiou, J. Y. (1978). Leaf protein concentrate as a protein source in diets for carp and rainbow trout. *Bull. Jap. Soc. Sci. Fish.*, **44**, 49.

Ohshima, M. (1985). The second limiting amino acid of Ladino clover LPC in rats. *Jap. J. Zootech. Sci.*, **56**, 267.

Ohshima, M. & Kogure, K. (1984). Factors affecting the quality of silages prepared from fibrous residues left after the extraction of leaf protein concentrates. *J. Jap. Soc. Grassld Sci.*, **30**, 178.

Ohshima, M. & Moriyama, Y. (1985). Effect of ethanol treatment of heat-coagulated LPC before oven-drying on the nutritive value in rats. *Jap. J. Zootech. Sci.*, **56**, 144.

Ohshima, M. & Oouchi, K. (1979). Ensiling characteristics of fibrous residues left after extraction of leaf protein concentrates from Ladino clover. *J. Jap. Grassld Soc.*, **25**, 260.

Ohshima, M. & Sogo, M. (1984). The digestibility of fibrous residues left after the extraction of leaf protein concentrate in cows. *J. Jap. Soc. Grassld Sci.*, **30**, 269.

Ohshima, M. & Ueda, H. (1982). Nutritive value of leaf protein concentrate and condensed green juice prepared from oats–annual meadow grass mixture for growing pigs. *Jap. J. Zootech. Soc.*, **53**, 622.

Ohshima, M. & Ueda, H. (1983). The nutritional evaluation of baker's yeast grown on oats leaf brown juice left after the separation of leaf protein concentrate in rats, chicks and pigs. *Jap. J. Zootech. Sci.*, **54**, 543.

Ohshima, M. & Ueda, H. (1984). A pilot green crop fractionation plant in Japan and the nutritive value of the products in pig and rat. In N. Singh (1984), p. 159.

Oke, O. L. (1966). The introduction of leaf protein into the Nigerian diet. *Nutrition*, **20**, 18.

Oke, O. L. (1983). Leaf protein research in Nigeria. In Telek & Graham (1983), p. 739.

Olatunbosun, D. A. (1976). Leaf protein for human use in Africa. *Indian J. Nutr. Dietet.*, **13**, 168.

Olatunbosun, D. A., Adadevoh, B. K. & Oke, O. L. (1972). Leaf protein: a new protein source for the management of protein calorie malnutrition in Nigeria. *Niger. med. J.*, **2**, 195.

Omole, T. A., Oke, O. L. & Mfon, B. P. (1976). Carrot leaf protein: preliminary trials with whole leaf using rabbits. *Nutr. Rep. Int.*, **14**, 173.

O'Neill, C., Jordan, P., Bhatt, T. & Newman, R. (1986). Silica and oesophageal cancer. In *Silicon biochemistry*, eds D. Evered & M. O'Connor. J. Wiley, Chichester, UK, p. 214.

Ornes, W. H. & Sutton, D. I. (1975). Removal of phosphorus from static sewage effluent by water hyacinth. *Hyacinth Contr. J.*, **13**, 56.

Osborne, T. B. (1924). *The vegetable proteins*, 2nd ed. Longmans, London.

Osborne, T. B. & Wakeman, A. J. (1920). The proteins of green leaves. I. Spinach leaves. *J. biol. Chem.*, **42**, 1.

Osborne, T. B., Wakeman, A. J. & Leavenworth, C. S. (1921). The proteins of the alfalfa plant. *J. biol. Chem.*, **49**, 63.

Ostrowski-Meissner, H. T. (1980*a*). Quantities and qualities of protein extracted from pasture herbage using heat precipitation or ultrafiltration procedures. *J. Sci. Fd Agric.*, **31**, 177.

Ostrowski-Meissner, H. T. (1980*b*). Protein degradation in herbage extraction during membrane filtration process: the effect of reducing agents addition. *J. Fd proces. Preserv.*, **4**, 261.

Ostrowski-Meissner, H. T. (1983*a*). Protein extraction from grasslands. In Telek & Graham, (1983), p. 9.

Ostrowski-Meissner, H. T. (1983*b*), Protein concentrates from pasture herbage and their fractionation into feed- and food-grade products. In Telek & Graham (1983), p. 437.

Ostrowski-Meissner, H. T., Carlsson, R. & Tragardh, C. (1980). Isolation and purification of protein from green vegetation for direct human consumption. In *Food process engineering*, eds P. Linko, Y. Maiki & J. Olkku, Applied Science Publishers, London.

Oyakawa, N., Orlandi, W. & Valente, E. O. L. (1968). The use of *Eichhornia crassipes* in the production of yeast, feeds and forages. Quoted from E. C. S. Little, *Handbook of utilization of aquatic plants*. FAO, Rome.

Pal, S. R., Mukherjee, D. & Sengupta, K. (1982). A brief review of agronomical work done on different oilseeds at the Pulses and Oilseeds Research Station, Berhampore, West Bengal. In *Pulses and oilseeds in West Bengal*, ed. D. K. Mukherji, p. 183. Directorate of Agriculture, West Bengal, India.

Pandley, R. K. (1983). Effect of leaf and flower removal on seed yield of lentil (*Lens esculenta* (L.). *J. agric. Sci. Camb.*, **100**, 493.

Paredes-Lopez, O. & Camagro, E. (1973). The use of alfalfa residual juice for production of single-cell protein. *Experientia*, **29**, 1233.

Parry, D. W., Hodson, M. J. & Sangster, A. S. (1984). Some recent advances in studies of silicon in higher plants. *Phil. Trans. R. Soc. B*, **304**, 537.

Patriquin, D. G. & Knowles, R. (1972). Nitrogen fixation in the rhizosphere of marine angiosperms. *Mar. Biol.*, **16**, 49.

Pereira, S. M. & Begum, A. (1968). Studies in the prevention of vitamin A deficiency. *Indian J. med. Res.*, **56**, 362.

Peterson, D. W. (1950). Some properties of a factor in alfalfa meal causing depression of growth in chicks. *J. biol. Chem.*, **183**, 647.

Peto, R., Doll, R., Buckley, J. D. & Sporn, M. B. (1981). Can dietary beta-carotene materially reduce human cancer rates? *Nature*, **290**, 201.

Pickle, C. S. & Caviness, C. E. (1984). Yield reduction from defoliation and plant cutoff of determinate and semideterminate soybean. *Agron. J.*, **76**, 474.

Pienazek, I., Rakowska, M. & Kunachowicz, H. (1975). The participation of methionine and cysteine in the formation of bonds resistant to the action of proteolytic enzymes in heated casein. *Br. J. Nutr.*, **34**, 163.

Pierpoint, W. S. (1959). Mitochondrial preparations from the leaves of tobacco (*Nicotiana tabacum*). *Biochem. J.*, **7**, 518.

Pierpoint, W. S. (1969). *o*-Quinones formed in plant extracts. Their reactions with amino acids and peptides. *Biochem. J.*, **112**, 609.

Pierpoint, W. S. (1971). Formation and behaviour of *o*-quinones in some processes of agricultural importance. *A. Rep. Rothamsted exp. Stn*, 1970, 199.

Pierpoint, W. S. (1983). Reactions of phenolic compounds with proteins, and their relevance to the production of leaf protein. In Telek & Graham (1983), p. 235.

Pieterse, A. H. (1978). The water hyacinth (*Eichhornia crassipes*) – a review. *Abs. trop. Agric.*, **4**, 9.

Pinstrup-Andersen, P. (1982). Introducing nutritional considerations into agricultural and rural development. *Fd Nutr. Bull.*, **4**(2), 33.

Pirie, A. (1983). Preventing blindness by marrying health to horticulture. In Roy (1983), p. 539.

– and Vitamin A deficiency and child blindness in the developing world. *Proc. Nutr. Soc.*, **42**, 53.

Pirie, N. W. (1942a). Direct use of leaf protein in human nutrition. *Chemy Ind.*, **61**, 45.

Pirie, N. W. (1942b). Green leaves as a source of proteins and other nutrients. *Nature*, **149**, 251.

Pirie, N. W. (1950). The isolation from normal tobacco leaves of nucleoproteins with some similarity to plant viruses. *Biochem. J.*, **47**, 614.

Pirie, N. W. (1951). The circumvention of waste. In *Four thousand million mouths*, eds. F. LeGros Clark & N. W. Pirie, p. 180. Oxford Univ. Press, Oxford.

Pirie, N. W. (1953). Large-scale production of edible protein from fresh leaves. *A. Rep. Rothamsted exp. Stn*, 1952, 173.

Pirie, N. W. (1955). Proteins. In *Modern methods of plant analysis*, eds. K. Paech & M. V. Tracey, vol. 4, p. 23. Springer, Heidelberg.

Pirie, N. W. (1957). Biochemical Engineering. *Research*, **10**, 29.

Pirie, N. W. (1958). Large-scale production of leaf protein. *A. Rep. Rothamsted exp. Stn*, 1957, 102.

Pirie, N. W. (1959a). Leaf proteins. *A. Rev. Pl. Physiol.*, **10**, 33.

Pirie, N. W. (1959b). The large-scale separation of fluids from fibrous pulps. *J. biochem. microbiol. Technol. Engng*, **1**, 13.

Pirie, N. W. (1959c). Large-scale production of leaf protein. *A. Rep. Rothamsted exp. Stn*, 1958, 93.

Pirie, N. W. (1960). Water hyacinth: a curse or a crop. *Nature*, **185**, 116.

Pirie, N. W. (1961). The disintegration of soft tissues in the absence of air. *J. agric. Engng Res.*, **6**, 142.

Pirie, N. W. (1962). Progress in biochemical engineering broadens our choice of crop plants. *Econ. Bot.*, **15**, 302.

Pirie, N. W. (1963). The selection and use of leafy crops as a source of protein for man. *Proc. 5th int. Congr. Biochem.*, **8**, 53.

Pirie, N. W. (1964a). The size of small organisms. *Proc. R. Soc.*, B, **160**, 149.

Pirie, N. W. (1964b). Large-scale production of leaf protein. *A. Rep. Rothamsted exp. Stn*, 1963, 96.

Pirie, N. W. (1964c). Freeze-drying, or drying by sublimation. In *Instrumental methods of experimental biology*, ed. D. W. Newman, p. 189. Macmillan, New York.

Pirie, N. W. (1966a). Improvements to machinery. *A. Rep. Rothamsted exp. Stn*, 1965, 106.

Pirie, N. W. (1966b). Fodder fractionation: an aspect of conservation. *Fertil. Feed. Stuffs J.*, **63**, 119.

Pirie, N. W. (1970a). Large-scale protein preparations. *A. Rep. Rothamsted exp. Stn*, 1969, 132.

Pirie, N. W. (1970b). Weeds are not all bad. *Ceres*, **3**(4), 31.

Pirie, N. W. ed. (1971a). *Leaf protein: its agronomy, preparation, quality and use.* Blackwell, Oxford.

Pirie, N. W. (1971b). Some obstacles to innovation. *Pugwash Newsl.*, **9**, 16.

Pirie, N. W. (1972). The direction of beneficial nutritional change. *Ecol. Fd Nutr.*, **1**, 279.

Pirie, N. W. (1975a). The potentialities of leafy vegetables and forages as food protein sources. *Baroda J. Nutr.*, **2**, 43.

Pirie, N. W. (1975b). Some obstacles to eliminating famine. *Proc. Nutr. Soc.*, **34**, 181.

Pirie, N. W. (1976a). *Food resources: conventional and novel.* Penguin Books, Harmondsworth, UK.

Pirie, N. W. (1976b). Food protein sources. *Phil. Trans. R. Soc.*, B, **274**, 489.

Pirie, N. W. (1976c). Restoring esteem for leafy vegetables. *Appropriate Technol.*, 3(3), 24.

Pirie, N. W. (1977). The extended use of fractionation processes. *Phil. Trans. R. Soc.*, B., **281**, 139.

Pirie, N. W. (1980). The temporary preservation of leaf protein. *Indian J. Nutr. Dietet.*, **17**, 349.

Pirie, N. W. (1981a). Would food and health services benefit from a more intimate union? *Ecol. Fd Nutr.*, **10**, 237.

Pirie, N. W. (1981b). The need for more information about vegetables. In *Vegetable Productivity*, ed. C. R. W. Spedding, p. 6 Macmillan, London.

Pirie, N. W. (1982). Rational choice among food sources. In *Biology, society and choice*, eds J. P. Hudson & T. Cavalier-Smith, p. 9 Inst. Biol.

– and Realistic approaches to Third World food supplies. *Third World Planning Rev.*, **4**, 31.

Pirie, N. W. (1983). The place of leaf protein in a food program. In Roy (1983), p. 453.

Pirie, N. W. (1984a). Stability of β carotene in moist, preserved leaf protein. In N. Singh (1984), p. 263.

Pirie, N. W. (1984b). Novel sources of protein. *Proc. 6th Int. Congr. Fd Sci. Technol.*, **5**, 155.

Pirie, N. W. (1984c). Fluctuating food fashions. *Interdisc. Sci. Rev.*, **9**, 149.

Pirie, N. W. (1984d). Factors affecting β carotene destruction in moist preserved leaf protein. *Qual. Plant., Plant Fds Hum. Nutr.*, **34**, 229.

Pirie, N. W. (1985a). Comments on the importance of green vegetables. *Qual. Plant., Plant Fds Hum. Nutr.*, **35**, 73.

Pirie, N. W. (1985b). Extracted leaf protein in British agriculture. *Nature*, **315**, 720.

– and Re-discovering the leaf. *Milling*, **169** (12), 15.

Pirie, N. W. (1986). Some preliminary observations on treatments which increase and decrease the stability of beta carotene in preserved leaf protein. *Qual. Plant., Plant Fds Hum. Nutr.* (in press).

Pirie, N. W., Fairclough, D. & Shardlow, A. W. (1958). Improvements in apparatus for the expulsion of fluid from fibrous materials. *Br. Pat.* 800 778.

Pleshkov, B. P. & Fowden, L. (1959). Amino-acid composition of the proteins of barley leaves in relation to the mineral nutrition and age of plants. *Nature*, **183**, 1445.

Powling, W. T. (1953). Protein extraction from green crops. *Wld Crops*, **5**, 63.

Prasad, V. L., Dev, D. V., Patil, R. E., Joshi, A. L. & Rangnekar, D. V. (1977). A note on feeding of lucerne extract to preruminant calves as part of milk replacer. *Indian J. Dairy Sci.*, **30**, 154.

Prokop, V., Gasnárek, Z., Kumprecht, I. & Jakobe, P. (1984). Lucerne protein-vitamin concentrate in pig fattening. *Živoč. Výr.*, **29**, 535.

Protein Advisory Group (1970). Statement on leaf protein concentrate. *PAG Bull.*, **10**, 3.

Proust, J. L. (1803). An essay on the fecula of green plants. *Phil. Mag.*, **16**, 122 and **17**, 22.

Ragster, L. E. & Chrispeels, M. J. (1981). Autodigestion in crude extracts of soybean leaves and isolated chloroplasts as a measure of proteolytic activity. *Pl. Physiol.*, **67**, 104.

Ramappa, B. S., Gowda, G. D. & Seshadri, T. S. (1986). Effect of inclusion of parthenium (*Parthenium hysterophorus*) meal in broiler diets. In Tasaki (1986), p. 213.

Rambourg, J. C. & Montes, B. (1983). Determination of polyphenolic compounds in leaf protein concentrates of lucerne and their effect on nutritional value. *Qual. Plant., Plant Fds Hum. Nutr.*, **33**, 169.

Randolph, J. W., Rivera-Brenes, L., Winfree, J. P. & Green, V. E. (1958). Mechanical dewatering as a potential means for improving the supply of quality animal feeds in the tropics and sub-tropics. *Proc. Soil Crop Sci. Soc. Florida*, **18**, 97.

Rangeekar, D. V., Patil, B. R., Prasad, V. L. & Joshi, A. L. (1979). Raising crossbred calves (Holstein × Gir) on milk replacer based on lucerne extract. *Indian vet. J.*, **56**, 306.

Rao, C. N. & Rao, B. S. (1970). Absorption of dietary carotenes in human subjects. *Am. J. clin. Nutr.*, **23**, 105.

Rawate, P. D. & Hill, R. M. (1985). Extraction of a high-protein isolate from Jerusalem artichoke (*Helianthus tuberosus*) tops and evaluation of its nutrition potential. *J. agric. Fd Chem.*, **33**, 29.

Raymond, W. F. (1977). Farm wastes. *Biologist*, **24**, 80.

Raymond, W. F. & Harris, C. E. (1957). The value of the fibrous residue from leaf-protein extraction as a feeding stuff for ruminants. *J. Br. Grassld Soc.*, **12**, 166.

Ream, H. W., Smith, D. & Walgenbach, R. P. (1977). Effects of deproteinized alfalfa juice applied to alfalfa-bromegrass, bromegrass and corn. *Agron. J.*, **69**, 685.

Ream, H. W., Jorgensen, N. A., Koegel, R. G. & Bruhn, H. D. (1983). On-farm forage harvesting: plant juice protein production system in a humid temperate climate. In Telek & Graham (1983), p. 467.

Reddy, G. U., & Joshi, K. G. (1984). Separation of chloroplastic, cytoplasmic fractions from leaf extracts by differential heat-coagulation and n-butanol precipitation. In N. Singh (1984), p. 181.

Řezníček, R. & Truxová, D. (1984). Physical methods of determining the degree of disintegration of lucerne plants. In *Physical properties of agricultural materials*, ed. R. Řezníček, translator J. Tauer, p. 127. Prague.

Richardson, M. (1977). The proteinase inhibitors of plants and microorganisms. *Phytochemistry*, **16**, 159.

Ries, S. K., Wert, V., Sweely, C. C. & Leavitt, R. A. (1977). Triacontanol: a new naturally occurring plant growth regulator. *Science*, **195**, 1339.

Roberts, E. A. H. (1959). The interaction of flavonol orthoquinones with cysteine and glutathione. *Chemy Ind.*, 995.

Rodriguez, E., Towers, G. H. N. & Mitchell. J. C. (1977). Allergic contact dermatitis and sesquiterpene lactones. *Compositae Newsl.*, **4**, 4.

Roja, F. C. & Smith, R. A. (1977). The antibacterial screening of some common ornamental plants. *Econ. Bot.*, **31**, 28.

Romero, A. J. R. & Diaz, A. C. (1984). A leaf protein concentrate from plantain (*Musa paradisiaca* L., subsp. *Normalis* O. Kze). In N. Singh (1984), p. 273.

Rothschild, M., Valadon, G. & Mummery, R. (1977). Carotenoids of the pupae of the Large White butterfly (*Pieris brassicae*) and the Small White butterfly (*Pieris rapae*). *J. Zool.*, **181**, 323.

Rouelle, H. M. (1773). Sur les fécules ou parties vertes des Plantes, & sur la matière glutineuse ou végéto-animale. *J. Méd. Chir. Pharm.*, **40**, 59.

Roy, S. K., ed. (1983). *Frontiers of research in agriculture.* Indian Statistical Institute, Calcutta.

Sale, P. J. M. (1973). Productivity of vegetable crops in a region of high solar input. I. Growth and development of the potato. *Aust. J. agric. Res.*, **24**, 733 & 751.

Sarathchandra, S. U. & Boyd, C. N. (1980), A microbiological study of the extracts of leaf proteins. *N.Z. J. agric. Res.*, **23**, 497.

Saunders, R. M., Connor, M. A., Booth, A. N., Bickoff, E. M. & Kohler, G. O. (1973). Measurement of digestibility of alfalfa protein concentrates by *in vivo* and *in vitro* methods. *J. Nutr.*, **103**, 530.

Savangikar, V. A. (1986). Identifying standard reference price of leaf protein concentrate. In Tasaki (1986), p. 221.

Savangikar, V. A. & Joshi, R. N. (1976). Influence of irrigation and fertiliser on the yields of extracted protein from lucerne. *Forage Res.*, **2**, 125.

Savangikar, V. A. & Joshi, R. N. (1978). Edible protein from *Parthenium hysterophorus* L. *Expl Agric.*, **14**, 93.

Savangikar, V. A. & Joshi, R. N. (1979). Modification of leaf protein concentrate by the use of plastein reaction. *J. Sci. Fd Agric.*, **30**, 899.

Savangikar, C. V., Savangikar, V. A. & Joshi, R. N. (1985). Fractional coagulation of proteins from alfalfa leaf juice by use of alum. *Proc. Indian Acad. Sci. (Plant Sci.)*, **95**, 47.

Schoney, R. A. & McGuckin, J. T. (1983). Economics of the wet fractionation system of alfalfa harvesting. *Am. J. agric. Econ.*, **65**, 38.

Schnabel, C. F. (1938). Vitaminic product from grass juice. *US Pat.* 2 133 362.

Schubert, K. R. & Evans, H. J. (1976). Hydrogen evolution: a major factor affecting the efficiency of nitrogen fixation in nodulated symbionts. *Proc. Natn. Acad. Sci. U.S.A.*, **73**, 1207.

Schwarz, K. (1977). Silicon, fibre, and atherosclerosis. *Lancet*, **i**, 454.

Scrimshaw, N. S. (1976). Strengths and weaknesses of the committee approach. *New Engl. J. Med.*, **294**, 136 & 198.

Sentheshanmuganathan, S. & Durand, S. (1969). Isolation and composition of proteins from leaves of plants grown in Ceylon. *J. Sci. Fd Agric.*, **20**, 603.

Shah, F. H. (1968). Changes in leaf protein lipids *in vitro*. *J. Sci. Fd Agric.*, **19**, 199.

Shah, F. H. (1971). Effect of heat on the extractability of lipid from leaf protein meal. *Pakist. J. scient. ind. Res.*, **14**, 492.

Shah, F. H. (1983). The future of leaf protein concentrate in Pakistan. In Telek & Graham (1983), p. 760.

Shah, F. H., Riaz-ud-Din & Salam, A. (1967). Effect of heat on the digestibility of leaf proteins. I. Toxicity of the lipids and their oxidation products. *Pakist. J. scient. ind. Res.*, **10**, 39.

Shah, F. H., Zia-ur-Rehman & Mahmud, B. A. (1976). Effect of extraction techniques on the extraction of protein from *Trifolium alexandrinum*. *Pakist. J. scient. ind. Res.*, **19**, 39.

Shah, F. H., Sheikh, A. S., Farrukh, N. & Rasool, A. (1981). A comparison of leaf protein concentrate fortified dishes and milk as supplements for children with nutritionally inadequate diets. *Qual. Plant.*, *Plant, Fds Hum. Nutr.*, **30**, 245.

Shearer, G. J. (1986). Commercial protein fractionation from lucerne. In Tasaki (1986), p. 196.

Sheen, S. J. (1983). Biomass and chemical composition of tobacco plants under high density growth. *Beit. Tabakforschung Internat.*, **12**, 35.

Sheen, S. J. & Sheen, V. L. (1985). Functional properties of fraction 1 protein from tobacco leaf. *J. agric. Fd Chem.*, **33**, 79.

Shurpalekar, K. S., Singh, N. & Sundaravalli, O. E. (1969). Nutritive value of leaf protein from lucerne (*Medicago sativa*): growth responses in rats at different protein levels and to supplementation with lysine and/or methionine. *Indian J. exp. Biol.*, **7**, 279.

Singh, G. (1984). Nutritional evaluation of some leaf protein preparations. In N. Singh (1984), p. 301.

Singh, G. & Singh, N. (1980). Fractionation of leaf proteins of lucerne (*Medicago sativa*) under low pH – low temperature treatment. *J. Fd Sci. Technol.*, **17**, 280.

Singh, N. (1960). Differences in the nature of nitrogen precipitated by various methods from wheat leaf extracts. *Biochim. biophys. Acta*, **45**, 422.

Singh, N. (1962). Proteolytic activity of leaf extracts. *J. Sci. Fd Agric.*, **13**, 325.

Singh, N. (1964). Leaf protein extraction from some plants of northern India. *J. Fd Sci. Technol.*, **1**, 37.

Singh, N. ed. (1984). *Progress in leaf protein research*. Today & Tomorrow's Printers & Publishers, New Delhi, India.

Siren, G. (1973). Protein ur skogsträd. *Svensk Naturv.*, **41**.

Siren, G., Blombäck, B. & Alden, T. (1970). *Proteins in forest tree leaves. R. Coll. For. (Sweden) Res. Notes*, **28**,.

Skinner, F. A. (1955). Antibiotics. In *Modern methods of plant analysis*, ed. K. Paech & M. V. Tracey, vol. 3, p. 626. Springer, Heidelberg.

Skorobogatykh, N. G. & Aitova, M. D. (1984). Green protein concentrate in feed mixtures for early-weaned piglets. Quoted from *Nutr. Abs. Rev. (B)*, **55**, 407, 1985.

Slade, R. E. (1937). Grass and the national food supply. In *Br. Ass. a. Rep.*, p. 457. Br. Ass. Adv. Sci.

Slade, R. E., Birkinshaw, J. H. & ICI (1939). Improvements in or related to the utilization of grass and other green crops. *Br. Pat.* 511 525.

Slade, R. E., Branscombe, D. J. & McGowan, J. C. (1945). Protein extraction. *Chemy Ind.*, **23**, 194.

Slansky, F. & Feeny, P. (1977). Stabilization of the rate of nitrogen accumulation by larvae of the cabbage butterfly on wild and cultivated food-plants. *Ecology Monographs*, **47**, 209.

Smiles, D. E. (1970). A theory of constant pressure filtration. *Chem. Eng. Sci.*, **25**, 985.

Smith, D. A. & Woodruff, M. F. A. (1951). *Deficiency diseases in Japanese prison camps. Med. Res. Coun. Spec. Rep.*, **274**.

Smith, R. C. (1972). Acetylation of methionine sulfoxide and methionine sulfone by the rat. *Biochim. biophys. Acta*, **261**, 304.

Sommer, A., Tarwotjo, I., Hussaini, G., Susanto., D. & Soegiharto, T. (1981). Incidence, prevalence, and scale of blinding malnutrition. *Lancet*, i, 1407.

Spencer, R. R., Mottola, A. C., Bickoff, E. M., Clark, J. P. & Kohler, G. O. (1971). The PRO-XAN process: the design and evaluation of a pilot plant system for the coagulation and separation of the leaf protein from alfalfa juice. *J. agric. Fd Chem.*, **19**, 504.

Squibb, R. L., Mendez, J., Guzman, M. A. & Scrimshaw, N. S. (1954). Ramie; a high protein forage crop for tropical areas. *J. Br. Grassld Soc.*, **9**, 313.

Srivastava, G. P., Tripathi, R. C. & Prasad, R. N. (1984). Tuber yield and extracted leaf protein yield from some potato varieties dehaulmed or harvested at different growth stages. In N. Singh (1984), p. 83.

Staron, T. (1975). Une méthode d'obtention des protéines à partir des plantes vertes. *C.R. hebd. Séanc. Acad. Agric. FR.*, **61**, 446.

Stockdale, C. R., King, K. R. & McKenzie, D. R. (1981). Nutritive value for lactating dairy cows of the pressed herbage remaining after the partial extraction of leaf juice. *Aust. J. exp. Agric. Anim. Husb.*, **21**, 376.

Straub, R. J., Tung, J. Y., Koegel, R. G. & Bruhn, H. D. (1979). Drum drying of plant juice protein concentrates. *Trans. Am. Soc. agric. Eng.* **22**, 484.

Strazetelski, J., Rys, R. & Lipiarska, E. (1981). Suitability of pulp from green feed in the diet of fattening bulls. *Acta Agr. Silv. Zootech.*, **20**, 213.

Subba Rao, M. S., Singh, N. & Prasanappa, G. (1967). Preservation of wet leaf protein concentrates. *J. Sci. Fd Agric.*, **18**, 295.

Subba Rau, B. H. & Singh, N. (1970). Studies on nutritive value of leaf protein from lucerne (*Medicago sativa*): II. Effect of processing conditions. *Indian J. exp. Biol.*, **8**, 34.

Subba Rau, B. H. & Singh, N. (1971). Studies on nutritive value of leaf protein from lucerne (*Medicago sativa*): III. Supplementation of rat diets based on wheat. *J. Sci. Fd Agric.*, **22**, 569.

Subba Rau, B. H., Mahadeviah, S. & Singh, N. (1969). Nutritional studies on whole-extract coagulated leaf protein and fractionated chloroplastic and cytoplasmic proteins from lucerne (*Medicago sativa*). *J. Sci. Fd Agric.*, **20**, 355.

Subba Rau, B. H., Ramana, K. V. R. & Singh, N. (1972). Studies on nutritive value of leaf proteins and some factors affecting their quality. *J. Sci. Fd Agric.*, **23**, 233.

Sullivan, J. T. (1943). Protein concentrates from grasses. *Science*, **98**, 363.

Sullivan, J. T. (1944). High-protein concentrate can be obtained from grass. *Fd Inds*, 186.

Sur, B. K. (1961). Nutritive value of lucerne-leaf proteins. Biological value of lucerne proteins and their supplementary relations to rice proteins measured by balance and rat growth methods. *Br. J. Nutr.*, 15, 419.

Swaminathan, M. S. (1981). Introducing nutritional considerations into agricultural and rural development. *Fd Nutr. Bull.*, 3(3), 30.

Synge, R. L. M. (1975). Interactions of polyphenols with proteins in plants and plant products. *Qualitas Pl. Mater. veg.*, 24, 337.

Synge, R. L. M. (1976). Damage to nutritional value of plant proteins by chemical reactions during storage and processing. *Qualitas Pl. Mater. veg.*, 26, 9.

Tapper, B. A., Lohrey, E., Hove, E. L. & Allison, R. M. (1975). Photosensitivity from chlorophyll-derived pigments. *J. Sci. Fd Agric.*, 26, 277.

Tarwotjo, L., Sommer, A., Soegiharto, T., Susanto, D. & Muhilal (1982). Dietary practices and xerophthalmia among Indonesian children. *Am. J. Clin. Nutr.*, 35, 574.

Tasaki, I., ed. (1986). *Recent advances in leaf protein research*. The proceedings of the Second International Conference on Leaf Protein Research. Iroha Publishing Co., Nagoya, Japan.

Taylor, D. L. (1968). Chloroplasts as symbiotic organelles in the digestive gland of *Elysia viridis* (Gastropoda: Opisthobranchia). *J. mar. biol. Ass.*, 48, 1.

Taylor, K. G., Bates, R. P. & Robbins, R. C. (1971). Extraction of protein from water hyacinth. *Hyacinth Contr. J.*, 9, 20.

Tekale, N. S. & Joshi, R. N. (1976). Extractable protein from by-product vegetation of some cole and root crops. *Ann. appl. Biol.*, 82, 155.

Tekale, N. S. & Joshi, R. N. (1977). Studies on the stability of carotenoid pigments in lucerne vegetation and juice. *Indian J. Nutr. Dietet.*, 14, 161.

Telek, L. & Graham, H. D., eds. (1983). *Leaf protein concentrates*. AVI Publishing Co., Westport, Conn, USA.

Telek, L. & Martin, F. W. (1983). Tropical plants for leaf protein concentrates. In Telek & Graham (1983), p. 81.

Terapuntuwat, S. & Tasaki, I. (1984). Nutritive value of leaf protein concentrates for young chicks; Protein and amino acid digestibility, biological value of protein and energy metabolizability of some leaf protein concentrates in chickens; and Protein utilization and amino acid digestibility of some leaf protein concentrates supplemented with amino acids in chickens. *Jap. Poultry Sci.*, 21, pp. 20, 65 & 89.

Thakur, M. L., Somaroo, B. H. & Grant, W. F. (1974). The phenolic constituents from leaves of *Manihot esculenta*. *Can. J. Bot.*, 52, 2381.

Thresh, J. M. (1956). Some effects of tannic acid and of leaf extracts which contain tannins on the infectivity of tobacco mosaic and tobacco necrosis viruses. *Ann. appl. Biol.*, 44, 608.

Thung, T. H. & van der Want, J. P. H. (1951). Viruses and tannins. *Tijdschr. PlZiekt.*, 57, 72.

Toosy, R. Z. & Shah, F. H. (1974). Leaf protein concentrate in human diet. *Pakist. J. scient. ind. Res.*, 17, 40.

Tracey, M. V. (1948). Leaf protease of tobacco and other plants. *Biochem. J.*, **42**, 281.

Tragardh, C. (1974). Production of leaf protein concentrate for human consumption by isopropanol treatment. A comparison between untreated raw juice and raw juice concentrated by evaporation and ultrafiltration. *Lebensm. wiss. Technol.*, **7**, 199.

Trench, R. K., Boyle, J. E. & Smith, D. D. (1973). The association between chloroplasts of *Codium fragile* and the mollusc *Elysia viridis*. I. Characteristics of isolated *Codium* chloroplasts. *Proc. R. Soc.*, B, **184**, 51.

Trigg, T. E. & Topps, J. H. (1971). The effects of additional methionine on the quality of leaf protein concentrates differing in nutritive value. *Proc. Nutr. Soc.*, **31**, 45A.

Tso, T. C. & Kung, S. D. (1983). Soluble proteins in tobacco and their potential use. In Telek & Graham (1983), p. 117.

Tsuchihashi, M. (1923). Zur Kenntnis der Blutkatalase. *Biochem. Z.*, **140**, 63.

Ueda, H. & Ohshima, M. (1983). Nutritive evaluation of leaf protein concentrates made from different crops in chicks. *Jap. Poultry Sci.*, **20**, 284.

United Nations (1968). *International action to avert the impending protein crisis.* UN, New York.

United Nations University (1980). *Nutritional evaluation of protein food.* (Eds P. L. Pellett & V. R. Young.) Tokyo.

United States President's Science Advisory Committee (1967). *The world food problem.* Govt. Printing Off., Washington.

Van Sumere, C. F., Albrecht, J., Dedonder, A., de Pooter, H. & Pé, I. (1975). Plant proteins and phenolics. In *The chemistry and biochemistry of plant proteins*, eds. J. Harborne & C. F. Van Sumere, p. 211. Academic Press, London.

Vartha, E. W. & Allison, R. M. (1973). Extractable protein from 'Grasslands Tama' Westerwolds ryegrass. *N.Z. J. exp. Agric.*, **1**, 239.

Vartha, E. W., Fletcher, L. R. & Allison, R. M. (1973). Protein-extracted herbage for sheep feeding *N.Z. J. exp. Agric.*, **1**, 171.

Venkataswamy, G., Krishnamurthy, K. A., Chandra, P. & Pirie, A. (1976). A nutrition rehabilitation centre for children with xerophthalmia. *Lancet*, **i**, 1120.

Vickery, H. B. (1945). The proteins of plants. *Physiol. Rev.*, **25**, 347.

Vickery, H. B. (1956). Thomas Burr Osborne. *J. Nutr.*, **59**, 3.

Vithayathil, P. J. & Murthy, G. S. (1972). New reactions of *o*-benzo-quinone at the thioether group of methionine. *Nature New Biol.*, **236**, 101.

Walker, A. F. (1982). The estimation of protein quality. In *Developments in food proteins*, ed. B. J. F. Hudson, vol. 2, p. 293. Applied Science Publishers, London.

Wallace, G. M. ed. (1975). *Leaf protein concentrates (New Zealand scene).* Publ. Ruakura agric. Res. Centre, Ruakura, N.Z.

Walsh, K. A. & Hauge, S. M. (1953). Carotene: factors affecting destruction in alfalfa. *J. agric. Fd Chem.*, **1**, 1001.

Walter, W. M., Purcell, A. E. & McCollum, G. K. (1978). Laboratory preparation of a protein-xanthophyll concentrate from sweet potato leaves. *J. agric. Fd Chem.*, **26**, 1222.

Wang, J. C. & Kinsella, J. E. (1976a). Functional properties of novel proteins: alfalfa leaf protein. *J. Fd Sci.*, **41**, 286.

Wang, J. C. & Kinsella, J. E. (1976b). Functional properties of alfalfa leaf protein: foaming. *J. Fd Sci.*, **41**, 488.

Wareing, P. F. & Allen, E. J. (1977). Physiological aspects of crop choice. *Phil. Trans. R. Soc.*, B, **281**, 107.

Waterlow, J. C. (1962). The absorption and retention of nitrogen from leaf protein by infants recovering from malnutrition. *Br. J. Nutr.*, **16**, 531.

White, J. R., Weil, L., Naghski, J., Della Monica, E. S. & Willaman, J. J. (1948). Protoplasts from plant materials. *Ind. engng Chem.*, **40**, 293.

Whitehead, R. G. (1980). Animal models for the study of protein-energy malnutrition. *Proc. Nutr. Soc.*, **39**, 227.

Widdowson, F. V. (1974). Results from experiments measuring the residues of nitrogen fertilizer given by sugar beet, and of ploughed-in sugar beet tops, on the yield of following barley. *J. agric. Sci. Camb.*, **83**, 415.

Wieringa, G. W. (1983). Aspects of green crop fractionation in relation to the utilization of grass. In Griffiths & Maguire (1983), p. 27.

Wilkins, R. J. ed. (1977). *Green crop fractionation.* Occasional Symposia 9, Br. Grassld Soc., Maidenhead, UK.

Wilkins, R. J., Heath, S. B., Roberts, W. P. & Foxell, P. R. (1977). A theoretical economic analysis of systems of green crop fractionation. In Wilkins (1977), p. 131.

Willaman, J. J. & Eskew, R. K. (1948). *Preparation and use of leaf meals from vegetable wastes. USDA Tech. Bull.*, **958**.

Williams, A. P., Hewitt, D. & Cockburn, J. E. (1979). A collaborative study on the determination of cyst(e)ine in feeding stuffs. *J. Sci. Fd Agric.*, **30**, 469.

Williams, W. (1979). Studies on crop yields – experiments at Reading University. In *Maximizing yields of crops.* H.M. Stationary Office (UK).

Wilson, R. F. & Tilley, J. M. A. (1965). Amino acid composition of lucerne and of lucerne and grass protein preparations. *J. Sci. Fd Agric.*, **16**, 173.

Winterstein, E. (1901). Ueber die stickstoffhaltigen Bestandtheile grüner Blätter. *Ber. dt. bot. Ges.*, **19**, 326.

Withers, N. J. (1973). Production of kenaf under temperate conditions. *N.Z. J. exp. Agric.*, **1**, 253.

Witt, S. C., Spencer, R. R., Bickoff, E. M. & Kohler, G. O. (1971). Carotenoid storage stability in drum dried Pro-Xan. *J. agric. Fd Chem.*, **19**, 162.

Wong, E. (1973). Plant phenolics. In *Chemistry and biochemistry of herbage*, eds. G. W. Butler & R. W. Bailey, vol. 1, p. 265. Academic Press, London.

Woodham, A. A. (1971). The use of animal tests for the evaluation of leaf protein concentrates. In Pirie (1971a), p. 115.

Woodham, A. A. (1983). The nutritional evaluation of leaf protein concentrates. In Telek & Graham (1983), p. 415.

Woodham-Smith, C. (1962). *The great hunger. Ireland 1845–1849.* Hamish Hamilton, London.

Wooten, J. W. & Dodd, J. D. (1976). Growth of water hyacinths in treated sewage effluent. *Econ. Bot.*, **30**, 29.

Worgan, J. T. & Wilkins, R. J. (1977). The utilization of deproteinised forage juice. In Wilkins (1977), p. 119.

WHO/U.S.AID (1976). *Vitamin A deficiency and xerophthalmia. Tech. Rep. Ser.*, **590**. WHO, Geneva.

Yamashita, M., Arai, S. & Fujimaki, M. (1976). Plastein reaction for food protein improvement. *J. agric. Fd Chem.*, **24**, 1100.

Yang, S. F. (1970). Sulfoxide formation from methionine or its sulfide analogs during aerobic oxidation of sulfite. *Biochemistry*, **9**, 5008.

Yemm, E. W. (1937). Respiration of barley plants. 3. Protein catabolism of starving leaves. *Proc. R. Soc., B*, **123**, 243.

Yemm, E. W. & Folkes, B. F. (1953). The amino acids of cytoplasmic and chloroplastic proteins of barley. *Biochem. J.*, **55**, 700.

Young, H. E. (1976). Muka: a good Russian idea. *J. For.*, **74**, 160.

Young, V. R. & Bier, D. M. (1981). Protein metabolism and nutritional state in man. *Proc. Nutr. Soc.*, **40**, 343.

Zimmermann, U., Pilwat, G. & Riemann, F. (1975). Preparation of erythrocyte ghosts by dielectric breakdown of the cell membrane. *Biochim. Biophys. Acta*, **375**, 209.

Zubrilin, A. A. (1963). In discussion in Pirie (1963).

Author Index

Note Page numbers in parentheses are locations of full references for papers cited in the text on the first-numbered page.

192

Author index

Subject index

Page numbers refer in *italic* to figures in text. Abbreviation LP used for leaf protein.

201

Subject index

crop species (*cont.*)
  water weeds, 24, 40–42
  weeds, 45–7
  see also individual species
Crotalaria juncea, 36
curry cubes, containing LP, 108–9
Cyamopsis psoralioides, 134
cyanohydrins, 86
cyst(e)ine, 62
  in acid hydrolysis, destruction of, 62,
    64, 65
  in amino acid analysis of LP, 62, 65
  modification of, in LP, 97–101
'cytoplasm' fraction of LP, digestibility
    of, 85
'cytoplasm' protein of LP, xi, 66–7
  acid-coagulation of, 50
  amino acid composition of, 63, 64,
    65
  digestibility of, 86, 87, 95
  methods of separating, 65–72
  nitrogen content of, 66–7, 85
  ultrafiltration of, concentration
    increase in, 71–2

Dactylis glomerata, 25, 26
dark green leafy vegetable (DGLV)
  β-carotene (pro-vitamin A) content,
    45, 123
  in vitamin A deficient diets, promotion
    of, 124
Daucus carota, 96, 130
deleterious components in LP, 102–4
'dewater' crops, 126
'dhal balls', containing LP, 110
digestibilty of LP
  of 'chloroplast' and 'cytoplasm'
    protein, 95
  drying temperature and, 77–8
  and heat-damaged LP, 87–9
  nitrogen and increased, 86
  preparative technique and improved,
    86
  in vitro, 84–9
  in vivo, 89–90; phenolic compounds
    and, 94–7; unsaturated fatty acids and,
    101
dilution, LP production and costs of,
    148
dog's mercury, *see Mercurialis perennis*
dry matter (DM) content
  average crop, distribution of, 125
  in commercially dried crop, 127
  of crop, fertiliser and, 21
  of crop, in period between harvests
    and, 28
  water removal and crop, 127
  in 'whey', 132

yield, and intervals between harvests,
    30
dry seed, harvested, 31
drying
  β-carotene (pro-vitamin A) loss, in LP,
    80–1
  energy costs of, 149
  of fibre, 127, 128, 130
  of LP, in air current, 77; freeze, 76–7,
  - main changes when heated during,
    88–9, in oven, 76, 77–8, by solvent
    extraction, 78–9
  unfractionated fodder, 149
dyes, possible by-products of LP, 136

Eichhornia crassipes, 11, 40–1, 93, 102,
    130
elder, *see Sambucus nigra*
Emblica officinalis, 88
Endomycopsis fibuliger, 'whey' as
    medium for, 134
energy
  in crop drying, 149
  in pulping, 145
  LP supplementation in diets and, 118,
    119, 120
enzymes
  heat coagulation and, 50, 51
  in digestive tract, 84–6
  on heat-damaged LP, 87–9
  in leaf extracts, 6–7
  on LP,
estrogens, in clover, 102–3
ethanol, 78, 135, 156
Eucalyptus saligna, 39
eutrophication, water weed growth and,
    40, 41
expressible, contrasted with extractable,
    protein 20, 23
extenders for LP, 77
extractability of LP
  age of leaves and, 5, 21, 26, 27
  alkali addition,
  increased, 41, 49
  intra-species variability and, 21–3
  with protein content, 21
  rubbing leaves and, 8
extraction units, mobile, where useful,
    41

fat hen, *see Chenopodium album*
fatty acids
  age of leaves, and pattern of, 58
  dominant in LP, 58
  inhibition of oxidation in, 88
  unsaturated, reactions of, 58, 101
fenugreek, *see Trigonella foenum-graecum*
fertiliser, nitrogenous

machinery for extraction and processing of
LP (*cont.*)
  hammer mill, 11–12
  high-speed pulpers, 145
  horn-angle press, 146
  IBP pulper and press, 12, 17–19, 20,
    28
  'pelleting press', 142
  power consumption, 10, 139, 149
  screw expeller, 145, twin, 126
  tests of different types, 10
Maillard reaction, 54, 62, 75
maize, *see Zea mays*
Manihot esculenta, 22, 32
Manihot utilissima, 32
manufacture of LP (pilot plant, 116–17
  quality in, 93
  two-stage pressing in, 128
  see also under fodder fractionation
Medicago sativa
  ash removed from LP of, 83
  chlorophyllase in, 50
  coumestrol in, 103
  cyst(e)ine in LP of, 65
  flavour of LP, 46
  fractionation of LP, 68
  LP, lysine shielded by sulfite in, 96
  LP yield of, 21, 26, 27–8, 29
  pheophorbide formation in, 50
  protease inhibitor, in LP of, 83, 84
  saponin in LP of, 22, 60
  'Simazine' on, 29
  Superfloc A 150 on, 71
  yeast yield, from 'whey' of, 135
Melilotus, 28
Mercurialis perennis, 46
methionine, 62
  conversion to sulfone and sulfoxide, 98
  modification of, 97–101
  structure of, 99
methionine sulfone, structure of, 99
methionine sulfoxide, structure of, 99
Mexico, LP production and use in,
  120–21
microbial culture medium, 'whey' as,
  134
microorganisms
  avoidance of LP contamination with,
    50, 52, 75
  'whey' as medium for, 134–5
mitochondria, 6
mobile extraction units, where useful, 41
Mucor racemosus, 75
mungo bean, *see Phaseolus mungo*
Musa paradisiaca, 37
Musa sapientum, 37
  and LP pie, 109
Musa textilis, 36

mustard, *see Sinapis alba*
mycotoxins, 60

nasturtium, *see Tropaeolum majus*
nettles, *see Urtica dioica*
Nicotiana tabacum, 37, 22
nicotine, conjugation with phenoloic
  substances, 61
Nigeria, LP feeding trial in, 116–17
Nile cabbage, *see Pistia stratiotes*
nitrogen, 3
  content of, in original crop and in
    extracted fibre, 7, 128–9
  protein and nonprotein, in
    trichloroacetic acid precipitation, 17
  ratio of nonprotein to protein, differs
    within and between species, 22
  in 'whey', 132
  in wild plant leaf often greater than in
    cultivated plant leaf, 43
nucleic acids
  destroyed by leaf ribonuclease, 17
  removal by delay between pulping and
    coagulation, 51
  in ribosomes, 6
nutritive value of LP
  animal diets, 91–3
  effects on, of modification of cyst(e)ine
    and methionine, 97–101
  extenders and, 77
  measurement of 112–23 *see also*
    children
  phenolic and other tanning agents on,
    94–7
  solvent extraction and, 79
  sulfite additions, effect on, 62
  unsaturated fatty acids, 58, 101
Nymphaea lotus, 42
Nymphaea odorata, 42

oats, *see Avena sativa*
oilseed, *see Brassica carinata*
oleic acid, 58
Oryza sativa, 31
oxalate
  in leafy vegetables, 44
  in sugar beet tops, 33
Paecilomyces, 135
palmitic acid, 58
palmitoleic acid, 58
Panicum maximum (grass), 30
papain, digestibility of LP by, 86–7
Parthenium hysterophorus, 38, 45, 104
pea, *see Pisum sativum*
pectin methyl esterase, pH of leaf pulps
  and, 49
penicillin, 'whey' as medium for, 134,
  135